On Time

On Time

Causality and the Quantum Gravity Conflict

JAN ZAANEN

OXFORD
UNIVERSITY PRESS

Great Clarendon Street, Oxford, OX2 6DP,
United Kingdom

Oxford University Press is a department of the University of Oxford.
It furthers the University's objective of excellence in research, scholarship,
and education by publishing worldwide. Oxford is a registered trade mark of
Oxford University Press in the UK and in certain other countries

Published in the United States of America by Oxford University Press
198 Madison Avenue, New York, NY 10016, United States of America

British Library Cataloguing in Publication Data
Data available

Library of Congress Control Number: 2024934122

ISBN 9780198920779

DOI: 10.1093/9780198920793.001.0001

Printed and bound by
CPI Group (UK) Ltd, Croydon, CR0 4YY

Preface

In a way the birth of this book has been a happy accident. Book projects typically involve quite a bit of hassle. One has for instance to figure out a good plan to seduce publishers to approve the project, especially when dealing with very credible ones like Oxford University Press. But this one sailed through in a rather playful, effortless fashion. The big deal is that I was diagnosed with oesophageal cancer in late summer 2022, an infamous killer. I went into an intense treatment mill, involving among others a five-week period of daily proton radiotherapy. This was followed by surgery in early 2023 that unfortunately went badly wrong, landing me in ICU in a coma for three weeks. Working full force on physics became difficult given the limits on my concentration span. However, a longstanding item on my bucket list has been to attempt to write a quasi-popular book. For a physicist, I appear to have some kind of writing talent—for my track record see the prologue. But the text I produced was always tied to a particular narrow physics affair and I dreamed about writing a text freely, without strings attached. Given my cancerous state I found that I had a right to pamper myself in this way, and I started writing, as a mere hedonist pursuit.

One, however, also needs something to write about. In the course of my long career as a theoretical physicist, I had collected original ways to conceptualize and explain the deep insights encoded in the fanciful mathematical machinery of modern physics. This has a reputation of being incomprehensible for anybody, with the exception of the mathematical nerds populating the theoretical physics floors. I may overestimate my powers in this regard, but I deemed it possible to catch this unrecognized, hidden affair in a non-technical (no equations) descriptive way, following the good example of, for instance, the various string theorists (Lennie Susskind, Brian Greene, ...), with their tradition of highlighting the remarkable workings of their very fancy mathematical achievements in this way.

Since I have been in recent times increasingly exposed to this string theory community, rather out of nowhere I got infected by a rather unusual outlook on the very basics of the number one problem of physics: the conflicting relationship between general relativity and quantum theory, the quantum–gravity problem. It is not that I have answers to offer, but I may guess something right by asking questions that to the best of my knowledge have never been asked before. The case that I have to offer is that the gravity–quantum conflict begins with the very different role of time in both theories. Time should be viewed in its most fundamental way. In the big world of gravity and humans, time is the domain of causality. Invariably, everything that happens in the future, is caused by events in the past. However, it is hard-wired in the structure of the quantum theory, governing the microscopic world and explaining the nature of matter, that 'when left alone' the time of quantum mechanics is strictly acausal. It is governed by the mathematical principle of 'unitary time evolution', that prohibits any form of information processing.

Finally, when the macroscopic and microscopic realities meet, the grand mystery called the collapse of the wavefunction kicks in. I will highlight the ideas introduced by the famous Penrose, that the collapse is also rooted in the conflicting roles of time in general relativity and quantum theory, where one may argue that gravity will destroy the unitary evolution on some mesoscopic scale. I will then review some very modern ideas of how this may relate to the stochastic nature of thermodynamics that governs the energy household in the big world, with its famous second law affair, insisting that useful energy always deteriorates in part in useless heat.

Somehow it seems I guessed something right, because in several iterations the story started to write itself: the remarkable swing of the story that emerged took me by surprise. Fast forward to late spring this year: untreatable metastatic growth has been detected in my brain—I will die in some near future. But I also got the manuscript finished, wondering what to do with it since in my perception it had spontaneously turned into a rather interesting text. I had been in touch with various book publishers in the past and I decided to just send an email to Sonke Adlung of Oxford University Press with the request to have a look at it. The remainder is now history—Sonke liked it very much and decided to publish it, going into overdrive to organize the further publishing process given the limited time I still have. I am very grateful for this greatly enjoyable course of events, cheering up my last days on the planet!

There are more friends that deserve a thank you. Various physics friends proofread the manuscript: Nico Pos, Dirk van der Marel, Jasper van Wezel, and Philip Phillips. I even managed to lure my brother-in-law Gerrit, an influential medical doctor, into giving it a try: to my impression he suffered quite a bit given his limited background knowledge in physics. This book is just too challenging for the layman, as you need to be at home with the basics of modern physics, albeit it is hopefully comprehensible also without having spent five years of your life on academic physics course work. There is one person that deserves special acknowledgement: Tjerk Oosterkamp. He is a colleague from Leiden, an experimental physics professor, making his living mostly by building instruments (see Fig. 4.2). He is some kind of unique hybrid, on the one hand a tinkering engineering person, but at the same time a genuine philosopher with an appetite for nearly impossible, game-changing experiments. He got infected with the Penrosian gravitational wavefunction collapse idea a long time ago, and all along I acted as his theoretical advisor in these matters. It will become clear from reading the text why I involved Tjerk quite a bit, among others asking him to have a look at early and crude versions of the story. Later he helped out with practical matters. I am especially grateful for his very thorough, final finebrush text edit before we submitted the final version to Oxford.

It remains to acknowledge the many people that have cared so well for my person during the agony of the disease. This includes a little army of medical folks—their empathy continues to astonish me—but of course especially those that I care most about: my dear wife Christa who has been my close companion for 35 years. And not to forget a very special oriental shorthair cat called Lola, that cheers up my life every night with her unrestrained affection.

Jan Zaanen, Leiden, October 2023

Contents

1

Prologue

The story that follows is about physics, and I am aiming at a most elementary affair. Despite the fact that we know so much regarding the workings of physical law, one often hears that more than ever before in history, we know what we do not know. This, in the first instance, revolves around the clash between the two grand theories of physics: general relativity describing the physics at the largest scales of the universe versus quantum theory governing the behaviours of the smallest parts. It has been recognized for a long time that under circumstances where both forms of physics meet, there are severe tensions, to the degree that seemingly the two theories appear to be mutually exclusive. This 'quantum-gravity' mystery is the main subject of this text.

I have no pretence to offer any answers. However, to stand a chance, the first crucial step is to ask the right questions. My career has been unusual, in the sense that I grew up being dedicated to the quantum physics cause, learning general relativity at a relatively old age. This appears to have triggered the effect that my brains got infected by some simple but quite unusual thoughts, that increasingly invaded my basic outlook on this matter. In fact, when listening to mainstream quantum-gravity stories I feel a strong urge to point out that these overlook the real problems. Although it is not solving any problem, I perceive these personal notions as strangely powerful. In the hands of somebody better than me, they may be helpful in guiding us towards answers.

Whatever it means, I would like to try to infect your brain, too, dear reader. These matters are in a way so elementary that I would like to try to cater with this text for a much broader readership then the usual highly specialized small crowd of condensed-matter physics experts, that have been the target of the hundreds of papers that I have authored in the past. 'In the land of the blind, one eye is king', in the physics community I do have a reputation for being responsible for appealing prose (see e.g. ref. [1])—I enjoy myself writing popular text, which is surely extra reason to engage on this mission impossible.

In modern street language a scientific credo is 'better first nail the no-brainers, to stand a chance with the brainers'. A case in point is the famous apple that gave Newton the idea to chase down his law of gravitation. No-brainers are about facts that are overly familiar, but somebody has to recognize the profundity hiding behind the ordinary. The no-brainers I have in the offering are perhaps best summarized by *it is time, stupid!*, paraphrasing a famous political quote [2]. In the realms of the

On Time. Jan Zaanen, Oxford University Press. © Oxford University Press (2024).
DOI: 10.1093/9780198920793.003.0001

fundamental theory, all the difficulties in my personal reference frame appear to hinge on the meaning and role of time.

What is time? In physics, it is a dimension like the dimensions of space. Since Einstein's relativity revolution we think in terms of *space time* having d spatial dimensions +1 time dimension: space time is $d + 1$ dimensional where d of the observable universe is three. But the time 'axis' is clearly different from the space dimensions. In space one can move freely in all directions – not quite true on our planet, but this is due to the gravitational interaction that makes it difficult to move in the vertical direction. Think in terms of the isotropic outer space where there is no difference between the x, y, and z directions. But in time the universe moves annoyingly only in one direction: from the past to the future. A very fundamental trait of the time direction as we experience it, is that it goes always hand in hand with *causality*, the fact that a cause at an earlier time leads to an effect at a later time. This is a necessary condition for the higher-level phenomenon called *information processing*.

With the arrival of the computers this 'act of computing' became increasingly the subject of a new branch of mathematics: mathematical information and complexity theory. It surely hangs together with causality: the output of a computation is clearly different from the input yet again indicative of the arrow of time: the output corresponds with 'better organized' information as compared to the input.

This has a mirror image in the physical world. A great achievement of nineteenth-century physics was the discovery of *thermodynamics* as motivated by the desire to optimize steam traction. Deeply rooted in the stochastic nature of the underlying microscopic physics one learns that useful work is always accompanied by the generation of heat, that will get lost since it cannot be converted back to work. The underlying principle revolves around *entropy*, a quantity that is actually a measure of the capacity to compute in information theory while it is a measure of probability as well. This has always to *increase* in the future direction: the *second law of thermodynamics*. In the physics literature it is often argued that this is the fundamental reason for time being characterized by the arrow.

Causality and information processing are taken for granted, these come to us as most basic traits of the universe. But are these completely self evident? In Chapter 2, I will present and fortify a claim that may sound at first quite provocative: *causality only exists in a universe with gravitation*. The arguments are very simple. I am completely relying on well-known results of general relativity and quantum theory, that are trustworthy since this is all within the domains of applicability of both theories. It is not unlike Newton's apple: It has just been overlooked but, at least in my head, it acts as a game changer.

Within the gravitational context the 'causality needs gravity' principle arises already on the elementary level of Newtonian gravity—for the physicist, it just hinges on the uniformly attractive nature of gravity: see Section 2.1. However, turning to General Relativity (GR), which explains the gravitational force in terms of the curved space times described by Riemannian differential geometry, causality is an intriguing manner on centre stage. Causality itself appears to be *hard wired* in the geometry itself! This intriguing affair—I find it myself the climax of GR—was only fully realized in the late 1960s with a key role by Roger Penrose, who received the 2020 Nobel prize for it.

The mind bender is the 'Penrose' or 'conformal' diagram, capturing the 'causal structure', an overview of the workings of cause and effect in a space time. The poster child is the Penrose diagram of the eternal black hole. This is a first physics secret that I will attempt to explain to the general reader. The trouble is that, in order to establish it, one needs to first suffer through years of maths training. But I am prejudiced that, as more often with GR, the outcomes can be explained in a descriptive language. The hardship is really in accepting how *strange* the world can be: a hostile reflex based on the desire to maintain psychological sanity may be unavoidable. But give the brain some time to get used to it and it may turn into the joy coming from enlightenment.

After preliminaries regarding GR in Sections 2.2–2.3, I will zoom in on the black hole Penrose diagram in Sections 2.4–2.5. This is all text book material and it does not contain any news for anybody who took a GR master course. Except then perhaps for the change in outlook. After realizing how crucial gravity is for the existence of causality, the way that Penrose diagrams reveal the intertwinement of geometry, force, and causality appears at least in my head as an intriguing witchcraft.

There is, however, yet another, very different, notion of time, which appears to be exclusively associated with the realms of nature governed by the laws of quantum physics. The take home message of this whole story is that this 'quantum time' and the 'causal time' of GR and human reality are intimately related but there is plenty of tension as well in this relationship. I interpret this as the signal of us missing fundamentals: it is the first question to address in attempting to solve the quantum gravity puzzle.

I like the phrase 'mathematical eyeglass': by suffering through the maths courses, one gets eventually rewarded as a physicist by a much wider view on reality, mediated by the equations. I can testify myself that by this mathematical training, somehow the brain twists are reconfigured and, in kind of a literal way, one can 'look into' this world hidden behind the equations. This appears to be what we mean by 'understanding physics'; I suspect that the visual system actually plays a crucial role.

Perhaps already pushing the limits, the Penrose diagram of Section 2.4 is a case in point. But for quantum physics, I have to push these limits much harder. The bottom line is that the quantum universe is so removed from our daily experiences that is impossible to appeal to intuition by employing effective metaphors. GR is here more human-friendly: the lively adventures in motion pictures based on accurate GR results such as the movie 'Interstellar' are testimony, and matters like time dilation near a black hole horizon are well explained. Such images are completely lacking in the quantum realms. But yet again, I cannot resist attempting to take the non-expert reader into the modern realms of quantum theory that are completely shrouded from the public eye, because there is a perception that these are unexplainable.

As a subplot, in the (lengthy) Section 3.1, I will engage in an experiment in this regard. These matters have the reputation to be so fanciful that only the specialized theorists of physics departments can handle it. I am prejudiced that it is actually *easier* to explain the essence of it to the mathematical blind. I believe I have some vivid images in the offering, capturing these essences. But it is also my favourite way to get the workings of time in quantum physics into focus—in this regard this may also be of interest to the physics readership. This way of viewing is not widely disseminated beyond the theory floor.

The way that quantum physics is taught is strongly influenced by history. It all started with the tiny systems of atomic physics where one copes with the quantum physics of at most a handful of microscopic objects, like the electrons that orbit around the atomic nucleus. This 'quantum mechanics' was captured using the 'canonical formalism', a highly abstract mathematical affair that revolves around 'operators' acting on the abstract 'Hilbert space'. There is nothing wrong with this, but one has to really learn the formalism to get anywhere, and this is like learning to multiply in primary school, when only knowing how to add up numbers.

But these are actually rather contrived, 'unnatural' systems. Quantum physics governs the behaviour of matter and matter is typically formed from *infinities* of microscopic constituents. The way that properties of matter evolve from the microscopic quantum physics is really in terms of *collective* effects implied by the latter: the 'quantum sociology' behind the nature of reality. This has been the subject of the mainstream of physics research since the late 1930s. In a long development that is not at all over yet, this turned into the art called 'quantum field theory', a mathematical machinery that governs at the same time not only high energy particle physics but also matters like the phenomenon of superconductivity found at very low temperatures.

The mathematical language that I will rely on is resting on the Feynman *path integrals*. But these rest in turn on a much older physics highlight, that is similarly hidden from the public eye. It is the statistical physics as formulated by the Austrian Boltzmann in the late nineteenth century. It is the classical physics version that answers questions like 'why do infinities of classical molecules decide to form together a gas, a liquid or a solid?' This turns out to be controlled by a stochasticity, the act of 'throwing dice'.

At least for the subclass of many-body quantum problems that are really understood, the path integral insists that one can just use Boltzmann's dice to compute states of *quantum* matter. The bottom line is that the theatre where this happens is no longer formed by space (as in the classical, literal Boltzmann incarnation) but instead by the aforementioned *space time*. This only works however when the system is in a state of *perfect equilibrium*. This means that the same conditions of temperature, pressure, whatever else one can imagine, are everywhere exactly the same. This is actually behind the claim that gravity is needed for causality: imposed by the quantum physical 'Eigenstate thermalization' principle discussed in Section 4.1, in the absence of gravity the fate of nature would be to eventually form *always* such a state of perfect equilibrium.

But in perfect equilibrium it is impossible to realize any *causal* evolution and the causal time as we know it loses its meaning. But what remains in terms of time, associated with the space time where quantum physics unfolds, is the intriguing part that I called before 'quantum time'. This is not very accurate since there is a lot of action associated with causal time in quantum physics. But this equilibrium time is somehow in a perfect harmony with the quantum principles, which is not all true dealing with the causal time.

The essence of this quantum equilibrium time is captured by a very simple, but also quite mysterious mathematical operation called the 'Wick rotation'. This is associated with what is called the *signature* of the space-time geometry. As explained in Section 2.2, the difference between the space dimensions and the time direction of space time, is that time is multiplied by the quantity $i = \sqrt{-1}$. The world is then automatically endowed with causal structure. However, one can as well take away this $\sqrt{-1}$ factor and this 'Euclidean' or 'imaginary' time behaves in the same way as space.

This has, among others, the ramification that this 'Euclidean' space-time looses completely causality structure. But this Euclidean space time is the theatre where the equilibrium quantum physics unfolds, and where the path integral turns into a Boltzmannian affair. The analogue of temperature in this 'thermal field theory' is associated with the strength of the quantum fluctuations. However, how to address the physical temperature in this Path integral representation of quantum physics? This is very intriguing: as it turns out, this Euclidean time axis is rolled up in a circle with a radius $R_\tau = \hbar/(k_B T)$ where $\hbar, k_B$ are Planck's and Boltzmann constants while T is the physical temperature. When temperature is low the time axis is long and when is temperature is high it is short. Botzmann's classical limit just corresponds with asserting that quantum time lasts for such a short interval that one can altogether ignore it and just focus on the space dimensions.

This highlights the a causal nature of the equilibrium state: by moving to the Euclidean future, one pops up in its own past since time is a circle. In addition, the only remnant of time under these equilibrium conditions is *temperature*: temperature is time-like, it sets the overall extent of this 'tranquility time'.

It is not at all the end of the time story in quantum physics. The big deal is, however, that the great triumphs of quantum physics in explaining the nature of reality are all found in this equilibrium corner—the subject of Section 3.1. We have here very powerful mathematical machinery at our disposal, while there is somehow a perfect harmony in the way that this 'pure' a causal quantum physics works. This is a bit tricky: with 'pure' I refer here to the so-called *unitary time evolution* 'sector' of the quantum story. This explains why quarks and gluons form neutrons and protons, that combine together with electrons into atoms, that in turn form molecules and eventually macroscopic phases of matter. The thermal field theory story that I just announced is machinery that is exclusively processing the unitary part. Equilibrium nature is the theatre of unitary quantum physics.

Surely, this 'Euclidean','thermal' quantum field theory affair is well known, it is the industry standard on the theory floor. However, I like to put it here in the limelight, because it is somehow maximizing the contrast between the time of gravity and the macroscopic world and the 'tranquility' (Euclidean) time that is the prevalent time in the microscopic quantum realms. The latter is often presented as a convenient mathematical trick, but I do not find that this does justice to this affair. It reveals somehow a deep insight—I do not know what it precisely means, but I am prejudiced that this contrast should be explained by a unifying quantum-gravity theory.

I am basically done announcing the content of Chapters 2 and 3. Hopefully, I will be successful in getting across the intrigue with the very different way that time works in gravity and in quantum physics, in that it is not forced to communicate with the *causal* nature of the macroscopic world. This stark contrast is just present in the equations that are time tested. My contribution is merely conceptual, just presenting it in an unusual order.

However, there is a part two to the quantum-time story that revolves around what happens when this quantum tranquility is forced to give in to the demands of the causality of the big world rooted in gravity. This is the subject of quantum dynamics and quantum measurement, eventually revolving around the collapse of the wave function. The discussion of this starts in Section 3.3, getting into full focus in Chapter 4. As with regard to the role of time, we realize that this measurement affair is about marrying the tranquil unitary time evolution of 'pure' quantum physics with the causal acts (measurements and so forth) in the macroscopic world. We do have working algorithms to compute matters accurately, but these depart from improvisations. That the foundations are shaky is brushed under the rug, in the run-of-the-mill quantum physics courses. However, having the eyes focussed on the central role of causality, it becomes obvious. *Anything causal in quantum physics is just inserted by hand!*

Much of the material in Chapter 4 is dedicated to a tour of relatively modern ideas regarding 'quantum measurements' that are yet again less well disseminated, even in the physics community. I will start explaining the eye opener called 'quantum thermalization', an insight that started to rise in the 1990s. The basic statement is that when a large quantum system is prepared in a highly excited non-equilibrium state it will in the ensuing time evolution eventually settle into an observable reality precisely corresponding with a *thermal* state where the excess energy is converted in a rise of the temperature.

This somehow hints at a relation between the *collapse* of the wavefunction and the principles that actually govern the real time evolutions in the macroscopic world. It is likely familiar to the reader that quantum physics leads to probabilities for 'observable' entities to occur—these actually arise *because* of the collapse of the wavefunction, there is nothing probabilistic in the realms of the unitary time evolution. On the other hand, in the macroscopic realms everything is governed by 'steam engine logic' in the form of thermodynamics. As I already argued, this is a stochastic affair expressed for instance by the 'second law' insisting that entropy is always increasing. It is a quite recent idea to link the stochasticity of thermodynamics to the fundamental stochasticity rooted in the wavefunction collapse. I will highlight in Section 4.2, a recent proposal involving repeated spontaneous wavefunction collapses that appears to make this connection utterly precise.

This brings me back to the main theme regarding the role of time but now in the wavefunction collapse context. This actually employs yet another motive associated with the quantum-gravity conflict. A main actor in this regard is yet again Roger Penrose, who already in the 1980s had forwarded the notion of 'objective' *gravitational* wavefunction collapse (Section 4.3). This is rooted in the fundamental problem that it is impossible to discern a global time 'direction' dealing with different mass distributions, that are implied by any quantum superposition. Penrose speculates that unitarity is for these reasons completely *destroyed* in the macroscopic realms where gravity may become sufficiently dominating.

In the final Sections 4.4 and 4.5 of that chapter, I endeavour in exploring an original idea, in so far I am aware. I address the question, 'could it be that the repeated spontaneous collapses underlying thermodynamics are actually rooted in the gravitational collapse mechanism?' I find this appealing, but it is far from a closed subject. As I will explain there are quite confusing loose ends. These passages are specifically aimed at the physicists—this is just describing a subject, begging for further research.

The remainder is in the first instance intended to be just entertaining. In Chapter 5, I will zoom in on the present mainstream in quantum gravity which is mostly embodied by the string theory community. First I will praise their main mathematical asset, the 'AdS/CFT correspondence'. This is a remarkable affair, making it possible to address very complicated 'quantum matter' issues, by actually employing the equations of general relativity!

This was initially conceived as an asset to shed light on the quantum gravity problem, but in this regard I am quite sceptical. I perceive it as a bit of a tragedy that, contrary to the folklore in the string theory community, the 'correspondence' has completely failed to deliver anything with regard to *quantum* gravity. I anticipate that I will surely seriously annoy the quantum gravitational readership with the remaining parts of this chapter. Since the discovery of the Hawking black hole radiation in the 1970s this became a central motive defining the field of black hole quantum physics, that in turn is closely tied to later developments like AdS/CFT. Two highlight themes that developed, are the 'firewall paradox' and even more so the 'information paradox'. With regard to both subjects I am quite sceptical, as I will explain in Sections 5.2 and 5.3, hopefully in a language that can be followed by all readers.

Finally, as a very light dessert, I have the epilogue in the offering, that is seriously tongue in cheek. Here I will in the first instance leave physics and turn to matters of (artificial) intelligence. In a story line that departs from Harari's book 'Sapiens', subsequently involving my cat Lola as well as ChatGP, I will arrive at a metaphysics due to the classic Greek Plato, insisting that mathematics is literally bigger than life. This is then brought back to physics in the form of an outrageous claim that the 'hippies that saved physics' in the 1970s may have guessed it right, that wave-function collapse may require intervention by a Taoist style consciousness. You are cordially invited to trash this ...

I hope you will enjoy this grand tour of the fundaments of very modern physics, presented by a tour guide having a somewhat unusual viewpoint on these matters!

2

Gravity as the cradle of cause and effect

Perhaps more than anything else in the history of science, Einstein's realization of the equivalence principle is *the* example of recognizing the profundity hidden in the seemingly self-evident daily reality of humans. The fact that gravitational and inertial mass are the same, was already fully realized by Newton in his *Principia Mathematica*, but it did not ring any further bell. It is very close to daily experience—jump out of an airplane and the parachutist starts to accelerate because of the planet's gravitational pull. That there is drag force that leads to a fixed terminal velocity is then a correction that had to be figured out, but the basic sensation of the gravitational pull as a potential hazard for the limbs is likely realized by all of us in the early stage of our lives, before we are even able to speak. But Einstein's genius was in the recognition of the non-obvious in the seemingly self-evident human daily experience. The equivalence principle was after all the crucial ingredient that forced the insight that gravity is actually not a force in the conventional sense, but instead reflecting the curved geometry of space time.

Causality is even more of a take-it-for-granted aspect of the natural world, as it is perceived by the human brain. That a cause leads to an effect is perhaps the first experience of all of us upon leaving our mother's womb. The baby's empty stomach is the beginning of a causal evolution—it triggers an alarm such that the baby starts crying, having the effect that the mother takes care that the stomach is filled with milk. From this point onwards, our human existence revolves around causality; it is encompassing our existence to such a degree that it is perhaps not completely self-evident.

One has to dig deep in the mathematical theories of physics to discern the tension rooted in causality, a tension that I perceive as the very essence of the quantum-gravity mystery. In Newton's mechanics, causality is a postulate. It revolves around equations of motion, departing from an initial state to which a force is applied (cause), having the effect that things start to move around, turning into something else (effect). But our present understanding of gravity goes a step further. It appears to offer an explanation *why* we are able to apply the forces of Newton. This is perhaps best captured by the classic wisdom that the cosmologies based on general relativity are, by default, non-stationary. Non-stationary means in GR that things are all the time happening as function of time, and the action along the time 'axis' is coincident with the notion of cause and effect.

On Time. Jan Zaanen, Oxford University Press. © Oxford University Press (2024).
DOI: 10.1093/9780198920793.003.0002

2.1 Playing God with Newton's constant

This seems to be not widely realized: gravity has an exquisite status as a necessary condition for the very existence of the causal evolution in our universe. A simple thought experiment invoking no more than Newtonian gravity gives this away already.

As an empirical input we need the LAMBDA-CDM standard model of big bang cosmology. This is famously confirmed by the cosmic microwave background, associated with the epoch where the universe expanded and cooled to the point that the primordial plasma recombined in neutral atoms, such that the universe became transparent to black body radiation. This CMB background is famously homogeneous—the establishment of tiny inhomogeneities as observed by a line up of increasingly sensitive satellites has been a main driver of the cosmology revolution that turned it, since the 1990s, into a high-precision quantitative science.

In devising physics thought experiments, you have the right to 'play God', tweaking conditions to become unphysical to see the contrast with the real world. Let's play God in LAMBDA-CDM. Let the early epoch big bang happen as usual but interfere at the moment in time just after the CMB decoupled, *by switching off Newton's gravitational constant.* Keep everything else we know unaffected: the workings of quantum physics, the standard model of high-energy physics in the form of a Yang–Mills quantum field theory and so forth. What would the universe look like?

The most rigorous way I know to address this question, is by invoking the full quantum description of reality as we presently understand it, which works just fine as long as we do not have to worry about gravity. In the present context, the key ingredient is the quantum principle of 'Eigenstate Thermalization'. I will come back to it at length, later in Section 4.1. Anticipating the outcome of this discussion, upon unleashing the standard rules of quantum physics on the time evolution of a *many-body* quantum system driven out of equilibrium, the generic destiny of such a system will be that, over long timescales, it will restore an impeccably *thermalized equilibrium state.*

Hence, after switching off Newton's constant in the freshly decoupled cosmos, this rule is insisting that the tiny non-equilibrium CMB inhomogeneities will rapidly anneal away, as a consequence of quantum thermalization. Matter and radiation in this universe will approach, after a short while, a perfect thermal equilibrium state. What will this look like? All there is, is a 'fog' that is perfectly the same everywhere, characterized by no more than a global temperature, pressure, and densities of various elementary particles.

Given this kind of universe, how to accomplish anything causal? Think like a physicist. In order to learn anything about reality you would like to do an experiment. But an experiment is a causal affair. You have to construct a machine that can give nature a hit, causing a response that you can follow in time, to find out its effect. But how to build *any* machine when the available building material is in the form of a perfectly homogeneous primordial fog? This is impossible, and it is fundamental: causality *disappears* in a universe that is in a perfect thermal equilibrium state. Without causality, there is no notion of time in the usual sense. It is a timeless world, a form of ultimate 'motionless' eternity of a most depressing kind for our human soul.

But now consider what happens in our actual universe, where there is no God switching off Newton's constant. Newtonian gravity is unique in the regard that it is uniformly attractive—all others forces are characterized by charges of opposite sign that attract each other, with the effect that neutral objects are formed, be it in the form of the electromagnetically neutral atoms or even the confined colourless baryons, formed from the colour charges of the quarks and gluons of QCD. However, a consequence of the uniformly attractive nature of gravity is that, according to the equations of motion of Newtonian gravity, the tiny inhomogeneities in the CMB get amplified in the course of time, leading to a clumping of the gas clouds in stars, galaxies, and eventually black holes, in an affair that continues to change its appearance in time. Locally, these clumps get so dense and hot, that the primordial hydrogen and helium atoms discover that they are actually in a metastable state. These nuclei start to fuse into bigger nuclei, culminating in the well-understood fate of stars as function of their mass. After a long stretch of time, these come to rest as white dwarfs, or either these explode as supernovae leaving behind neutron stars or black holes, spewing the heavier 'metals' into the gas clouds. The next generations of stars are formed accompanied by planets formed from this cosmic ash.

The highly non-equilibrium state of the Sun, generating copious amounts of free-energy by nuclear fusion, has the effect that the planets are irradiated by intense light, a form of low entropy energy. This impinges on the Earth, where it causes the physics on the surface to be in a highly excited non-equilibrium state. This causes the ever-changing weather patterns, but it is also responsible for perhaps the greatest non-equilibrium act in the universe. Resting on the laws of chemistry, forms of organization emerged through the simple principles of Darwinian evolution, becoming increasingly complex and organized, turning into what we call biological life. This culminates eventually, in the information-processing capacity of our brain, perhaps the most extreme form of non-equilibrium realized in nature. As a caveat, we may find ourselves in an era where the machines that we have engineered with our big brains, may surpass this natural incarnation. Our computers famously produce enormous amounts of heat as a prize that one cannot avoid when keeping systems in an extreme non-equilibrium condition.

The take-home message is a no-brainer. This seemingly self-evident experience that all of our existence revolves around causality is not entirely self-evident. It requires a particular form of physical law: it needs *gravity*.

In fact, this 'intertwinement' of force and cause and effect becomes even more staggering in the theory resting on Riemannian geometry, of which Newtonian gravity is a special limit: general relativity. This already raises its head in non-gravitating *special* relativity. Next to the funny effects of time dilation and Lorentz contraction, a major consequence of which is the emergence of *causality* structure. It is just impossible to know what is going on elsewhere, when the light cone of elsewhere is not intersecting with your own light cone.

2.2   The square root of minus one and the origin of causality

Let me attempt to give the non-physicist a sense of how the mathematics behind the hard wiring of this causality in the geometry works. This is just physics text-book material [3] and I hope that I will succeed in getting it across in a very descriptive fashion. The physicist may want to glance through what follows, as I will perhaps put the emphasis a bit differently from the standard in the textbooks.

Minkowski realized that this can be captured by simple geometry. We are used to the geometry of the 'spatial manifold' which is endowed with the so-called Euclidean signature. This just means that the distance between two points is governed by Pythagoras' theorem. The infinitesimal distance (metric) ds is governed by $ds^2 = dx^2 + dy^2 + dz^2$ using Cartesian coordinates. As a caveat, notice that such a coordinate choice is by principle just a convenient way to label the manifold. The 'diffeomorphism invariance' or 'general covariance' is just insisting that the reality of space cannot depend on any particular choice of coordinate system. When the time τ would be like space, space-time should have the 'Euclidean signature' metric $ds^2 = c^2 d\tau^2 + dx^2 + dy^2 + dz^2$ postulating that there is an absolute speed limit set by the light velocity c. But the time of the physical universe is different. Instead, it is an *empirical* fact that the distance between two points in space time is set by $ds^2 = -c^2 dt^2 + dx^2 + dy^2 + dz^2$. There is a minus in front of dt^2: this is called the Lorentzian signature because the Lorentz transformations capturing the essence of special relativity just follow from this simple assertion.

A way to capture this conveniently, is by introducing complex numbers, through the trick of calling $\sqrt{-1} = i$, and writing numbers in the form of $a = b + ic$. It follows that $i^2 = -1$. Departing from the Pythagorean expression for the metric in Euclidean signature, by just replacing $\tau \to it$ (the 'Wick rotation') one infers that the time factor in the metric transforms as $c^2 d\tau^2 \to -c^2 d^2t$, and special relativity just drops out. This simple mathematical trick has staggering consequences in dealing with Einstein's view on the nature of space time. In fact, by invoking this i in the geometry, causality is *imposed* on the reality that it describes. I perceive it as perhaps the grandest mathematical miracle associated with the nature of reality.

2.3 Riemannian differential geometry as the backbone of general relativity

Physics is eventually about mobilizing the eerie powers of mathematics, to shed an otherwise counterintuitive light on the nature of the world—the 'mathematical eyeglass'. In this regard, I continue to be greatly impressed by general relativity: it mobilizes the highlight of nineteenth-century mathematics—Riemannian differential geometry—to shed a penetrating light on the history of the cosmos itself. Among physics students, differential geometry has kind of a bad reputation as being hard to learn. In most physics departments, the GR course, which amounts mostly to learning Riemannian geometry, is offered only at the graduate level.

Einstein himself suffered quite a bit for a number of years, trying to get a grip on Riemannian geometry, after being tipped off by his mathematican friend Grossman that it could fit his equivalence principle. Being not a real mathematician myself, I don't hesitate to admit that I also struggled with it, when I started to learn it when I was in my early forties. Being a student in the 1970s, I did not take the course back then, since only the overly mathematically inclined physics students took it. In the period 1930–1970, GR actually disappeared more or less from the physics portfolio, in fact being kept alive by the mathematicians.

Riemannian geometry is a strangely powerful affair. I am presently enjoying a theoretical machine, that is running on supercomputers, with an engine room stitched together from Riemannian parts, telling stories about the quantum nature of matter. I will come back to this 'holographic duality' machine later. But it highlights the magical powers of the Riemannian way. It is like dealing with the hardware in a successful experimental lab—I have never before in my rich career encountered a theoretical machine revealing such a richness, flexibility, and above all 'correctness', that is always right and trustworthy, in this regard.

Riemannian geometry, in its nineteenth-century incarnation, is dealing with translating the geometry of curved manifolds into algebraic equations that can be solved by applying mathematical-style rules. It generalized the musings of Gauss and others, who dealt with the two-dimensional curved surfaces of three-dimensional objects, matters like it being faster to fly over northern Canada on the way from Europe to California (see Fig. 2.1). But the essence of Riemann is that in three and higher dimensions, there is much more structure to this curved geometry than in 2D as revealed by the tensor calculus, an art that one has to learn in a similar way as one learns to count in kindergarten. This in turn is the secret to the propulsion system of GR: the Einstein equations of motion. It is all about these tensors: it states that the energy-stress tensor capturing the role of matter and energy 'is equal to' the Ricci tensor capturing curvature properties of the space-time where it all happens.

This backbone is very beautiful in its simple essence. However, in the general situation it is a system of highly non-linear partial differential equations, which is a mathematical way of saying that it is in a way not all that different from the equations governing the weather. Even with the latest generation of supercomputers one can only predict what will happen up to a week from now, before one gets completely lost in the myriad of numbers. However, dealing with highly symmetrical situations, the theory simplifies to simple differential equations that can be solved on the back of an envelope. These solutions form the backbone portfolio of GR, like the Friedman–Le Maitre–Robertson–Walker metric underlying the large scale structure of the cosmos. Another highlight are the black hole metrics: Schwarzschild, Reissner–Nordstrom, and the much more recent Kerr metric of the astrophysical rotating black holes, found by Kerr as recently as 1964.

But now the signature issue intervenes, the $i = \sqrt{-1}$ affair I alluded to in the above. The nineteenth-century maths tradition was surely pre-occupied with the Euclidean signature, the extra dimensions behaving just like space dimensions. But physical space-time is characterized by the Lorentzian signature, time being the outlier knowing about the i in front of it. This innocent looking 'Wick rotation' factor has, however, staggering consequences, that I find are discussed in an overly technical way in the GR textbooks. It is a deeply conceptual affair that tends to be presented as the outcome of a mechanical computation.

The Einstein equations are indifferent regarding signature, they work in both ways and the Lorentzian signature is an additional physics postulate. In Euclidean signature there is no sense of causality, in the same way that causality has nothing to do with the shortest geodesic between Amsterdam and San Francisco. However, in Lorentzian signature the curved geometry gets automatically endowed with a hard-wired, all-encompassing causality structure!

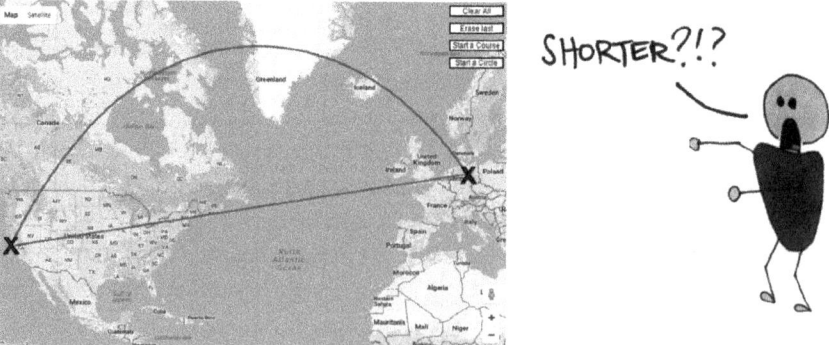

Fig. 2.1 An example of a counter-intuitive ramification of a curved geometry is that the shortest flight path between Europe and California over the curved surface of the Earth hits northern Canada.

2.4 Penrose diagrams and the causal structure of the eternal black hole geometry

As I stressed in the above, causality is already hard wired dealing with the flat space-times of special relativity. In the curved space-times of general relativity, endowed with Lorentzian signature, an objective (observer-independent) causality structure emerges that is put into full focus by the Penrose (better Penrose–Carter) diagrams. Penrose (Fig. 2.2) played a key role in the revival of GR in the late 1960s by his focussing in on this property of GR, that causality is an intrinsic property of the solutions of the Einstein equations. For instance, the singularity theorems that earned him a Nobel prize depart from a rather arbitrary mass distribution. However, given only these initial conditions one can identify a 'trapped surface': the causality hard-wired in the Lorentzian signature Einstein equations then insists that, upon waiting long enough, this matter will always fall in on itself, forming a black hole with its infamous singularity in the middle.

The Penrose diagrams are constructed by a special 'conformal' coordinate choice for the metric, pulling spatial and/or temporal infinities on a piece of paper, a mathematical highlight learned by any GR student. Let me again attempt to get the essence of the construction across to the non-physicists. This relies on simple geometrical diagrams that however represent conceptual hardship that any GR student has to overcome in *reading* these diagrams.

These are typically presented in the form of a two-dimensional diagram, one axis being a radial coordinate where every point represents the 'anchor' of a d-1

Fig. 2.2 A hero of this story, in more than one regard: Sir Roger Penrose.

dimensional 'slice' (think of a ball in 3D) of isotropic space. The other axis represents time, giving an overview of the various destinies of any observer regardless of the choice of coordinates. The mathematical magic behind the construction of these diagrams is in the trick of 'pulling' spatial and temporal infinities onto a finite piece of paper. This employs a coordinate transformation resting on a mathematical entity that is especially since the COVID pandemic quite familiar to the general public: the exponential function. In the COVID context this has the meaning of the extremely rapid growth of infected people, departing from the simple ingredient that after a given interval of time the number of infections double. One then finds an evolution like $2 \to 4 \to 8 \to 16 \to \cdots$. This growth is extremely rapid and in no time the whole population is infected. Think about this whole population as infinity but one can encode this as well in the actual number of doublings, which is a much smaller number that can actually be taken to be quite finite. This corresponds with the transformed coordinates used in the Penrose diagrams—all what really matters is that one can keep track of what is going on at these infinities.

The magic of these diagrams cannot be better highlighted than by the most ubiquitous of all GR solutions: the Schwarzschild geometry associated with a (spatially) isotropic 'eternal' (time-independent) black hole in an asymptotically flat universe. Let me just show the outcome and explain how it works. 'Conformal' time and space correspond with the vertical and the horizontal axes. Let us depart from flat (zero-gravity) Minkowski space-time, Fig. (2.3). This corresponds with a triangle, where time runs from minus (i^-) to plus infinity (i^+), and space from

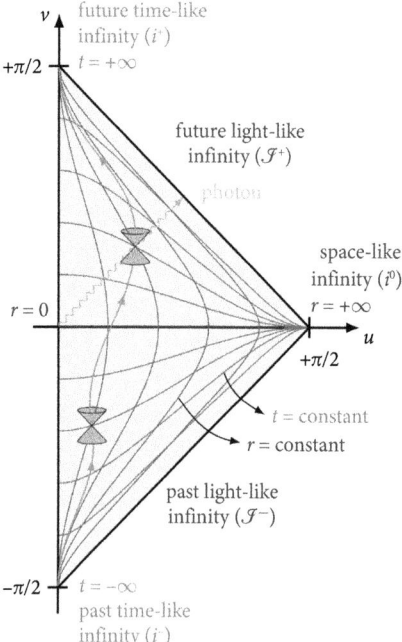

Fig. 2.3 The Penrose causality diagram of an infinitely large, flat 'Minkowski space-time'.

the left axis (the 'middle' of space, $R = 0$)) to spatial infinity i^0 (the right corner of the triangle) all pulled from infinity onto the piece of paper by the conformal transformation.

The crucial mathematical instrument that one needs to cope with dealing with causality in relativity is the light cone. This departs from the wisdom that there is an absolute speed limit—the velocity of light $c = 299{,}792{,}459$ metres per second. Metres and seconds are just human conventions, introduced by Napoleon, with velocity being the conversion factor. For something speeding with c, 1 second is the same thing as approximately 300,000 kilometres. In a first step, physicists use 'natural units' putting $c = 1$; at the end of the computations one can always restore the kilometres and seconds. Hence, in these natural units on a diagram where time increases from bottom to top and space from left to right this forms a cone with a 45° opening angle. Moving slower than light, which is of course admitted, means that you have to stay inside the light cone, while speeding as fast as a light ray means that you are seeing things located at the rim of the light cone, while it is impossible to be the cause of any effect outside the light cone. There is the causality: you can't get there and therefore you cannot influence it.

Back to the flat-space Penrose diagram. One reads off the causality, by drawing a light cone that is oriented according to the lines indicated in the figure: one sees immediately that an observer departing from minus temporal infinity (i^-) is 'causally connected' ('sees the whole diagram') to all of the space-time. On the other hand, the observer associated with the indicated light cone attached to the green line trajectory can surely get at the plus temporal infinity (i^+) but a large part of space to the left and the right of its cone is no longer in view. It is not in causal contact with this part of space-time. When you are not causality diagram literate, you should play a bit, moving light cones formed by your crossed fingers along the lines to get a better grip on the conceptual gymnastics of the workings of causality—this is entertaining when you get the gist of it.

This gets really interesting dealing with the causality structure of the Schwarzschild black hole, Fig. (2.4). Historically it was quite a discovery— although the solution was written down by Schwarzschild in terms of his coordinates as early as 1915, it took until 1960 for the mathematicians Kruskal and Szekeres to discover a coordinate system revealing *all* of the 'extended' space-time of this black hole. What follows just falls out of the mathematical solution—it is an example of the remarkable powers of Riemannian geometry.

Let's try to read the Schwarzschild Penrose diagram. There are 'diamond' squares sticking out to the left (III) and the right (I). Let us focus on the right one, region I. This encodes for an infinitely large universe outside the black hole. As for Minkowski, this starts from a time like minus infinity i^-. Divide this square into two triangles, and the triangle to the right is the same one as we just discussed for Minkowski space: this encodes for an infinitely large Minkowski, flat-space universe and you may think that you are somewhere in this region. However, there is also a left triangle and when you depart from i^- you can get there without

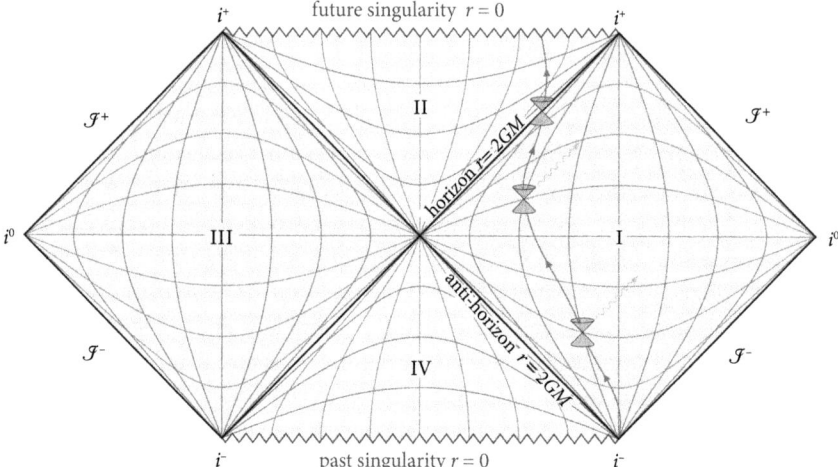

Fig. 2.4 The Penrose causality diagram of the extended space-time of the eternal Schwarzschild black hole.

violating the 'stay in the light cone' rule. This terminates at a full line diagonal called $r = 2GM$. G is Newton's constant and M the energy of the black hole that you have to input in the computation and this yields, in natural units, a length scale. This r is the famous Schwarzschild radius, telling us where the *event horizon* of the black hole resides.

Given the popularity of the black hole in popular culture this will ring a bell. This is the famous ghostly surface that is the 'point of no return': upon passing the horizon, the fate of even a light ray is sealed. It *has* to enter the interior of the black hole, and to keep out one needs to fly faster than light, which is not possible. You can see this directly from the diagram. Depart from i^- with a light cone with its opening looking straight up. Move a bit to the left, and continue to go straight to the horizon. At a point, the middle of your light cone will intersect with the horizon diagonal (you are passing the horizon) entering the black hole interior region II and you see that the right rim of your light cone coincides with the horizon diagonal. Move a bit further upward and it is no longer possible to get out from the interior II to the 'free universe' region I: you are locked up in the black hole interior region!

But this is just the first step. Look upwards in the interior region II and you see lingering in your future a *horizontal wiggly line* connecting your time-like infinity i^+ with another time-like infinity that I will explain in a moment. The ramification is that whatever you do, eventually you will *always* end up in your future at the wiggly line. Your fate is sealed, nothing you can do to avoid it!

This wiggly line encodes for a lot of effects: it denotes the *black hole singularity* that is, according to the equations, a point precisely in the middle of the black hole where the most extreme of all mathematical hells breaks loose. The equations explode in infinities; the most tangible of these infinities is associated with tidal forces. In the same way as the gravity of the Moon is responsible for the tides on Earth, upon getting close to the singularity, the gravitational force at the top of your head is much larger than at the tip of your toe, with the consequence that you get stretched out, getting longer and longer and thinner and thinner. This is called 'to be spaghettified'. Eventually you will look like a spaghetti with a diameter smaller than the radius of a proton. This is surely the most extreme fate one can imagine, being hard-wired in your future at the moment you pass the horizon. Appreciate the epic workings of cause and effect in Einstein theory!

But we have only deciphered part of the Penrose diagram. What else is going on? It is good to realize that what follows is special to the eternal black holes, the Schwarzschild solution where the black hole was always there in the past, while it will continue to be there in all of the future. Astrophysical black holes are of a different kind. These spring into existence at some point in time, by the gravitational collapse of a very heavy star nearing the end of its life, where it has burned up all its nuclear fuel. As far as we know, for such black holes the story in the previous paragraph captures cause and effect completely—the spaghettification affair, but no more than that. But eternal black holes are even more mind boggling!

Departing from i^- we only looked forward. But starting from i^- we can look to the left, finding yet another wiggly singularity horizontal line. However, you are causally disconnected from this part of space-time, you cannot get there from the right 'external universe'. But it is there. This 'bottom of the diagram' wiggly line is the 'time reversed' version of the future singularity at the top of the diagram: things can only *get out* of this so-called 'white hole' singularity. Departing from this bottom (wiggly line) singularity, travel again upwards towards the future and at a point you will cross the lower diagonal on the right side. These are things that are popping through the horizon from the interior region IV, entering our universe I. Now the rule is that these things can get out of the black hole interior behind this horizon, but they can never return. This 'lower' area is therefore called the 'white hole' since the course of events is here precisely opposite to the black hole further up the figure.

But this is not yet the whole story, because we have only explored the right half on the diagram. We see that the 'white hole' and 'black hole' diagonals cross in the middle while these lines continue on the left side of the diagram, producing a perfect mirror image region III of the 'our' universe I! On the left side there is a 'normal' universe as well, but now existing to the left of the horizon lines! This is called the two-sided black hole geometry: next to our ('right') universe there is a parallel ('left') universe, which is separated by the white/black hole interiors. There is, however, no way that we can traverse this black hole interior to get from our universe to the parallel universe without violating causality, as you may wish to check yourself with your crossed fingers.

You also see that the white hole interior, black hole interior and the two 'exterior' universes meet precisely at a point. This is called a 'non-traversable wormhole'. This relates to the wormholes used for warp drive purposes in science fiction movies, best depicted in the movie 'Interstellar'. Actually, the GR prophet Kip Thorne was co-producer—I know this first hand because Kip was in Leiden as visiting ('Lorentz') professor when this was in the making. He arranged that the Hollywood graphics people got help from the professional numerical GR effort, to add realism to the graphics (Fig. 2.5). I was surprised to learn that these graphics hackers were apparently based in London. For Kip it was convenient to be close by in Leiden!

Hence, the interstellar wormhole is the most realistic one ever seen on a movie screen, in the sense that the gravitational lensing effects responsible for the visual appearance are correct. But the difference with the eternal black hole wormhole is that this one is not traversable. It cannot be used to travel through it from the 'left' to the 'right' universe since this is causally forbidden. According to the present understanding of GR, it appears to be impossible to open up such wormholes so that a human can get through without violating physical laws.

Once again, isn't this a surprise of epic proportions, to find that the combinations of curved geometry and Lorentzian signature, processed by straightforward algebraic manipulations, reveals a world drenched in cause and effect? I perceived this myself as perhaps the most conceptually disturbing side of GR. I am prejudiced that it is to a degree brushed under the rug in GR text books, where it is presented as a 'technical' exercise illustrating the ways that one has to solve the equations. I have been teaching myself for a number of years the mid-level GR course. I could not resist presenting this two-sided black hole to the students as

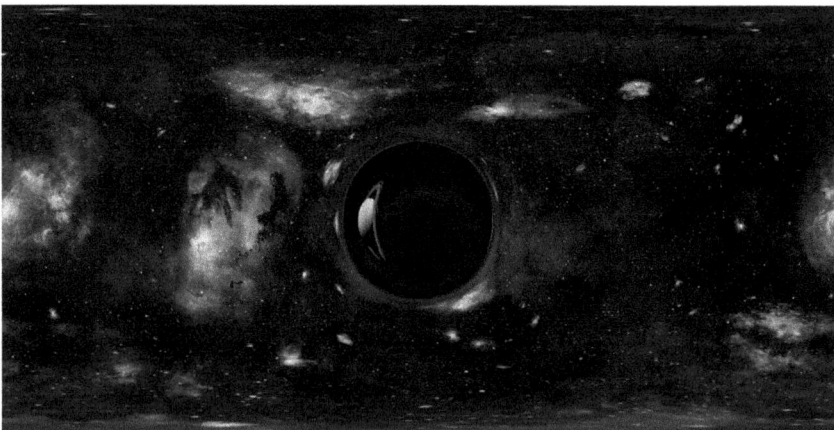

Fig. 2.5 The wormhole starring in the movie 'Interstellar', further processed at https://wzcorse.com/interstellar-wormhole for realistic gravitational lensing effects.

the climax of this whole affair, in a similar tone as in the above. But the student response was invariably disappointing; instead of the shining eyes that I expected after explaining this staggering insight, the students were looking at me as if it came from a different planet. It seems that it just takes quite some time for the human mind to get acquainted with this familiar (causality) but nevertheless greatly counterintuitive mathematical result. But this is the essence of all really good physics.

2.5 The black hole and Euclidean signature: the vanishing of cause and effect

I stressed that all this causality structure intrinsic to GR is eventually just originating in the Wick rotation: multiply the 'space-like' time in Euclidean signature with i and you are done. There seems to be no better way to illustrate this than to again focus in on the elementary Schwarzschild black hole. In fact, the Einstein equations don't care about signature. The signature of space-time is just an additional physics postulate: the theory is equally well behaved in both signatures albeit the solutions do look quite different. Again the Schwarzschild solution is a case in point. It is mathematically quite simple: write down the solution for the Lorentzian metric and substitute for the Lorentzian time the Euclidean version by $\tau = it$, and let's see what the equations will tell.

The famous man-in-the-wheelchair Stephen Hawking did obsess quite a bit about this 'Euclidean gravity', Einstein theory in Euclidean signature space-times. Together with Gibbons, he discovered the way that Euclidean Schwarzschild space-times look like. This is illustrated in Fig. (2.6). This is much simpler than the Lorentzian version of the geometry. All one needs to look at is what happens

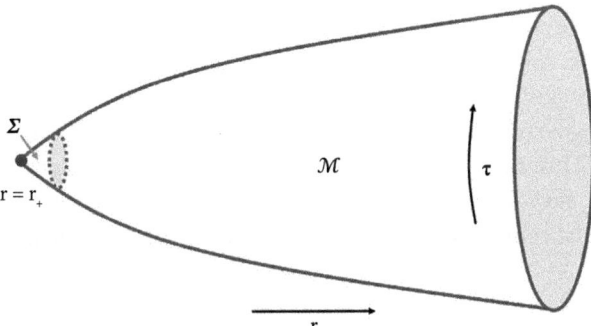

Fig. 2.6 The Schwarzschild black hole geometry as function of the radial coordinate r and imaginary time τ, given the Euclidean signature. r_+ is the location of the event horizon where the metric vanishes.

as a function of (radial) distance from the black hole r and (Eucidean) time τ. As I already stressed, causality disappears entirely when 'time is like space'. A first causal structure in Lorentzian signature is of course the event horizon separating the causally different interior and exterior of the black hole. But such knowledge cannot be encoded in Euclidean signature.

What is the maths telling us what is happening instead? It is very efficient: the interior region just does not exist in the geometry of the Euclidean black hole. The geometry just comes to an end at the horizon and only the analogue of the exterior geometry exists!

But the other outcome is even more striking: the Euclidean time coordinate τ is in this exterior geometry *rolled up in a circle*; the maths takes care that it is automatically 'compactified'. The radius of this time circle shrinks to zero at the horizon, and upon moving away from the black hole along the radial direction, this radius starts to grow to saturate at spatial infinity at a value set in natural units by again the combination $R_\tau \sim 8\pi GM$. The bigger the black hole, the larger this time radius.

The big deal is that this Euclidean is 'compact', it is a circle. The sense of the duration of time is the same for Lorentzian and Euclidean time, the difference being merely in the $\sqrt{-1}$ factor. So what happens departing at a certain time at spatial infinity, travelling to your 'Euclidean future'? Since this time lives on a circle when you travel in the forward time direction you will find that you pop up from your own past after a time R_τ! It seems that you have a time machine and you can travel in your future to your own past. But this is a deception. There is just no notion of cause and effect; literally nothing can happen in the course of time, and *therefore* you can 'circle around' in Euclidean time. The usual objection against time travel is rooted in causality, a famous illustration being the 'grandfather paradox'. When a time machine would exist that could bring you to the past, you could use the opportunity to kill your grandfather before your father was inseminated in your grandmother. But then you could not have been born, let alone having had the opportunity to kill grandpa.

But it is just completely impossible for such events to happen in Euclidean time. To convince yourself of this claim we have to first elucidate how time works in the theory that governs the way that material things behave, when gravity does not play a role. This is governed by quantum physics, where yet again, time is in the driver's seat, however, playing a role that is completely different, being actually in a fundamental conflict with the time of gravity that I just highlighted.

3

Quantum physics: the tranquillity
of the Euclidean time circle

Having arrived at this junction, I hope I have managed to get across the splendour of the way that causal evolution is hard-wired in the Lorentzian signature version of Riemannian geometry. Is there a stronger expression of the causal notion of fate than the sad fact that your destiny upon passing the event horizon will be in the form of being spaghettified at the singularity, while any attempt to avoid it is as futile as futile can be?

But how does the factor of time work in that other truly fundamental mathematical theory describing nature: *quantum physics*?

It is a very different affair. It is not at all the case that causal evolution is part of its mathematical machinery as it is in GR. In fact, quantum physics is celebrating its greatest successes in the rise of quantum field theory, taking most of the energy of the (quantum) theorists since the late 1930s. This became a great success story, especially so in high-energy particle physics, culminating in the standard model revolution of the 1970s. But what is less well known is that it also forms a backbone for the understanding of the way that macroscopic matter emerges from the microscopic quantum mechanics. This is not only about common forms of matter like solids, liquids, and gases, but also underlying the thorough understanding of 'quantum matter': superconductors and so forth.

This is the area where I have been working myself for most of my career. In the context of the role of time in physics, this quantum field theory tradition is however in a way strongly biased. That it is nearly entirely focussed on the 'timeless circumstances' of equilibrium physics, is something that I hopefully hit home effectively already in the previous chapter. In part this is for technical reasons: the available mathematical methodology is quite limited in its ability to capture non-equilibrium properties in the quantum field theoretical context. It is however magical to find out the surprisingly simple, elegant, harmonious, and efficient way of understanding the 'timeless time' equilibrium physics associated with quantum field theories. On purpose I will first take up this story. It at least imprinted in my brain a fear for the 'active' time of non-equilibrium, having the impression that this may well be a quite ubiquitous condition among my peers.

But surely there is a lot to learn regarding the working of quantum physics in the presence of causal evolutions. This was in fact the central subject in the

On Time. Jan Zaanen, Oxford University Press. © Oxford University Press (2024).
DOI: 10.1093/9780198920793.003.0003

very early days when Bohr and contemporaries were nearly entirely focussed on atomic physics experiments, dealing with the 'non-equilibrium' phenomena, associated with in fact rather contrived, tiny quantum mechanical systems. This is about the Copenhagen versus many world interpretation, the Schrödinger cat(s) (see Fig. 3.1), the Einstein–Podolsky–Rosen paradox, and so forth. These 'foundations' attracted a lot of attention among philosophers but much less so among real physicists. This tends to be in the foreground in popular accounts on quantum mechanics, but arguably much of it went round in circles in the 1930s.

But surely the recipes to deal with 'measurements', being at the centre of this 1930s affair, do deliver. In a very recent development it has been lifted to a new level by the rise of the technologically oriented 'second quantum revolution', aiming, among others, at the building of a quantum computer. But different from the 'timeless' field theoretical equilibrium quantum physics, I am of the strong opinion that this 'measurement' agenda involves improvisations and matters that are quite mysterious and beg for a better explanation—specifically the 'collapse of the wavefunction'.

To capture this contrast let me first present a short history of quantum field theory and the role of equilibrium time, before delving into the generalities associated with time in quantum physics.

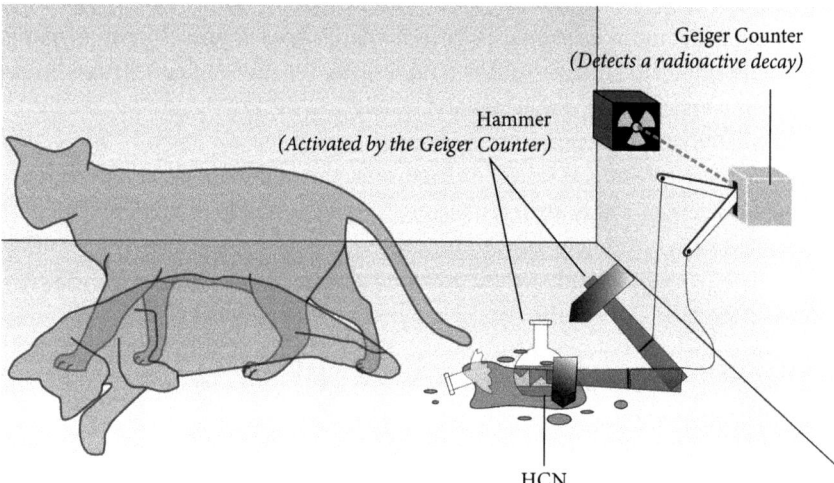

Fig. 3.1 I will assume that the reader is somewhat familiar with Schrödinger's cat, the symbol of quantum physics. Otherwise Google it.

3.1 A history of quantum physics: field theory as the big deal

This section will deal with the 'timeless' condition, in the sense of the Gedanken experiment cosmos of the big bang universe, with Newton's constant switched off, as I discussed in the above. As I argued, this universe is filled with the featureless 'fog', having everywhere the same temperature, pressure, and so forth. Causality has disappeared in such a cosmos since it is impossible to create circumstances that can act as causes, leading to effects. This is the 'equilibrium state'. It is a mental construction involving a Platonic perfection since in earthly circumstances it can only be approximately approached, given that our cosmos is inherently out of equilibrium due to the universal 'pull' of gravity.

However, it has been at the focus of attention of physics since the nineteenth century. This is due to the efforts of those who created thermodynamics in this era, having an eye on improving steam traction. This climaxed in the late nineteenth century by Boltzmann's formulation of classical statistical physics. A great deal of the hard work of the physics community since the 1930s has been in generalizing this into a fundamental and all-encompassing description of the nature of equilibrium matter, as we encounter it in the natural world in terms of the fundamental quantum theory.

In fact, many more physics papers have been published dealing with equilibrium physics than with the causal, non-equilibrium circumstances, for the simple reason that much more can be done: fully fledged quantum non-equilibrium dealing with the real physical world formed from infinities of microscopic building blocks is a technically hazardous affair. Presently, non-equilibrium quantum physics is regarded as a frontier motivated by progress with constructing various guises of quantum computers and simulators. This development is motivated by progress in the engineering of the required hardware, but when these machines start to buzz, great progress is expected in these realms.

What is the equilibrium quantum physics story about? It is the affair that is familiar to physicists because it is the core of the course program in physics departments. It does take a number of years of following maths-dense courses, before one can see this light in full. Given the need to expand the mathematical eye glass into a fully fledged mathematical telescope, this part of the physics narrative is much less disseminated in the public mind than for instance the science fiction friendly black hole and Schrödinger cat stories.

A disclaimer is in place here. I would dearly like to tell this story in a truly comprehensible way to the physics layman. However, the realities of quantum physics are detached to such a degree from our human world that it is a completely alien reality. I perceive it myself as a bit of a miracle that after years of training somehow the equations internalize—in one or the other way one is 'seeing' what it is about, and it becomes comfortable and pleasant. I got at a point that I find equilibrium quantum physics *easier* to grasp than our big world classical physics. But it appears to be a mission impossible to get this across to humans that have not suffered through the maths courses.

All I can do is to advertise what the physics community figured out during roughly a century of hard work, more or less starting with the accomplishments of Planck, Einstein, and Bohr at the beginning of the twentieth century, coming to a climax in the 1970s with the standard model of elementary particle physics. Eventually, the focus is on explaining the nature of matter and energy, departing from elementary constituents getting 'glued together', through the laws of quantum physics.

This started in the early twentieth century using the experimental information regarding *atoms* to find out how quantum *mechanics* works. It was understood that atoms contain a positively charged, heavy nucleus surrounded by a cloud of very light and therefore very quantal electrons. As a lucky circumstance, much of the way that these electrons do their quantum-job can be understood by considering a single electron moving in the electrical field of the nucleus. The quantum physics of one, or at most only a few things, is where it started and this subject area was called 'quantum mechanics'.

Nowadays a (superficial) description of this affair is taught, at least in Dutch high schools, and it may be familiar to the reader. Eventually, it boils down to the Schrödinger equation, insisting that elementary particles like electrons have a dual character: in quantum mechanics particles are also waves and the other way around, and what one gets to see depends on the question one asks. This has the ramification that the orbits of the electrons around the nucleus have to be *standing waves* that precisely fit the 'matter waves' of de Broglie. These are basically like the sound of a guitar string: the lowest-energy orbit is analogous to the fundamental mode, and the ladder of overtones correspond with the quantized higher energy states.

This is then to be combined with another fundamental postulate of quantum physics that rules when there is more than one particle in the system: quantum statistics. This is eerie and after all these years, I am still frightened by it. But we know it is at work because of experiment. It governs a basic rule of how quantum things *collectivize*. They do so by picking two extremes: they either want to all do the same thing—be in the same single particle quantum state—and such ultra-conformist particles are called *Bosons* (after the Indian physicist Bose). The only alternative for them, is to behave as *Fermions* (after the Italian master Fermi). Fermions are the ultra individualists: every fermion has to be in its unique quantum state, that is, different from all the quantum states 'occupied' by other particles. To make these matters even more confusing, such particles are actually *indistinguishable*: it is impossible to keep track of any one particle in particular.

Electrons are fermions and, accordingly, these have to occupy higher and higher quantized orbits when the atom number (counting the number of electrons) is increasing, moving to the right and down the periodic table. This is not yet the whole story, since the electrons repel each other mutually since they are equally charged. In fact, this is a problem that is not mathematically solvable, but approximation schemes—'perturbation theory'—were devised that yield relatively accurate outcomes. The qualitative take-home message is, that this 'shell model', where one ignores in first instance these interactions, yields a fairly accurate impression of atomic structure.

This story then continues with explaining the fundamentals of chemistry: how atoms combine in the plethora of molecules. The basics may also be familiar to the reader from high school. Atoms may decide to turn into charged ions by exchanging electrons, and then they are bonded together by the Coulomb force: the 'ionic bond'. But the other extreme is the covalent bond, where the electrons delocalize between the atoms. Since their 'effective box' becomes larger this takes less quantum kinetic energy, glueing together the neutral atoms. These are actually extremes and many real-life chemical compounds are somewhere in the middle. And there are as well more subtle quantum mechanical binding mechanisms such as the 'van der Waals' and 'hydrogen' bonding.

Given that molecules are by default larger than atoms, it becomes more difficult to compute their quantum properties quantitatively, especially to do justice to the effects of the electron–electron interactions. This turned into a huge effort involving supercomputers, called 'quantum chemistry'. Presently, this is part of the sales pitch for quantum computers: such machines are required, for instance, in order to compute the properties of molecules of pharmaceutical interest.

Although quantum chemistry is very hard work, its conceptual outlook is still best called quantum mechanical. It is eventually focussed on a restricted number of electrons orbiting a finite number of atoms. But macroscopic matter consists of near infinities of constituents: Avogadro's constant spells out that there are of the order of 10^{23} atoms per mole of matter. What is happening when such numbers approach infinity, dealing with the 'thermodynamic limit'? As far as I know, not much is explained in high school and this was actually for me the revelation taking the physics lecture courses in University. Once again, I anticipate that this may be the part that is quite unfamiliar to the non-physicists. I will do my best to get the guts of this insider story across.

This took off in the late nineteenth century, due to the brilliant achievements by Boltzmann laying the groundwork for what later became 'statistical physics'. In an earlier community effort motivated by the desire to improve steam traction, *thermodynamics* was discovered and developed. By itself an eerily powerful, mathematical affair, it explains the way that heat acts in macroscopic systems, introducing the notion of *entropy*. This boils down to the fact that upon processing energy resources, only part of it can be exploited for useful work, while one cannot avoid that a certain fraction gets unretrievably lost as heat. And that this energy cannot be recovered. This reveals the notorious 'second law' insisting that entropy is always increasing in the forward time direction. Remember this—it is the way that the fact that time always evolves to the future, and never to the past is understood in classical 'steam engine' physics. In Chapter 4, I will present the case of how this is eventually tied to the arrow of time in quantum physics.

Boltzmann related all of this: the microscopic physics of the colliding atoms and molecules forming the macroscopic gases and fluids. Back then, the very existence of atoms and molecules was a highly controversial affair, and the way that Boltzmann used the properties of the macroscopic gas to prove the existence of atoms is a mind-boggling achievement – in the early twentieth century it was a big deal for the young generation, including for instance Einstein, who added the decisive Brownian motion. The random motions of atoms, as directly exhibited by this Brownian motion, translate in the macroscopic realms as heat: the higher the temperature the more vigorous these motions. This has then the implication

Fig. 3.2 An unsung hero of physics: Ludwig Boltzmann, who did the groundwork in the nineteenth century for statistical physics. His tombstone in Vienna, indicating his famous formula for the entropy S in terms of the multiplicity of the micro-states W and his constant usually called k_B, B of course referring to Boltzmann.

that the higher temperature states are more *probable* and entropy is just a measure of this probability (see Fig. 3.2), enforcing the arrow of time through the second law.

But Boltzmann took this a step further, turning it into a quantitative framework resting on stochastic calculus. As a first step, these finite temperature 'classical' equilibrium states of matter are explicitly ignoring *time*. Departing from a classical mechanics description of the microscopic physics, Boltzmann mobilized the notion of *ergodicity*: in the course of time, the system can in principle explore all possible microscopic configurations. In order to address the equilibrium properties, one can just capture this by a probability distribution for the *static* configurations: the 'Boltzmann factor' insisting that the probability of a 'microstate' (configuration), i (p_i), is set by the exponential of minus the ratio of its potential energy E_i to temperature T: $p_i = \exp -E_i/(k_B T)$, where the Boltzmann constant k_B is just a conversion factor translating temperature units (centigrades, Kelvins) to energy units (Joules, electronvolts, etcetera).

I hope you are still with me after this first equation! To make matters worse in this particular regard, let me write Boltzmann's equation for the entropy S itself: $S = -k_B \sum_i p_i \ln p_i$, where $\sum_i$ means sum over all configurations i, and ln is the natural logarithm. To really confuse you, as it turns out, the central wheel in Bolzmann's machine room is the partition sum: $Z = \sum_i p_i$, the object to compute containing sufficient information to describe all observable equilibrium properties. Once again, what matters most for my particular purposes is that the thermal equilibrium states of classical physics are precisely captured by a theory that is lacking a notion of time altogether.

Departing from Boltzmann's fundamentals, in the century that followed, statistical physics turned into a triumphant affair explaining everything that is happening in the finite temperature equilibrium universe. The big deal is the explanation of the *phases* of matter, the most familiar being the solids, liquids, and gases. This landed on its feet, halfway through the twentieth century, with crucial contributions by the Soviet school under the leadership of Lev Landau, another unsung hero of physics (Fig. 3.3). It was discovered that macroscopic physics is often governed by the *order parameter field*, the poster child of the *emergence* principle. Single atoms and so forth have no knowledge regarding the question whether the system of atoms will form a solid, a gas or a liquid. However, infinities of them combine into a collective degree of freedom, that is determining whether the solid is formed or whether the system stays a liquid, pending macroscopic factors like pressure and temperature.

This is a remarkably elegant affair. Order parameter theory is tightly constrained by the most general ingredients: the symmetry governing the matter and the dimensionality of space. But this theory also describes a rich reality. The highlight in this regard is the theory of the *critical state*, realized at a (continuous) phase transition, as developed in the late 1960s and early 1970s. Right at the transition,

Fig. 3.3 Another unsung hero of physics: the Soviet Lev Landau, who should be considered, more than anybody else, the father of modern quantum field theory. Landau discovered the simple principles underlying the 'order parameters', the mathematical machinery underpinning our understanding of ordered, disordered, and critical states of matter. But his Landau school took this also all the way to the zero temperature quantum theories explaining the 'quantum liquids', such as the Fermi-liquids realized in metals, and the superconductors/superfluids. The famous 'Higgs mechanism' of high-energy physics is actually a trivial generalization of the Ginzburg–Landau theory describing superconductivity.

the two phases do coexist. Take the water—steam system and at a particular temperature and pressure ('critical end point'), one no longer finds the bubbles of steam, characterizing boiling water. Instead, at this 'continuous phase transition', one finds the bubbles to be distributed in a *scale invariant* way. You may be aware of fractals, also an expression of scale invariance: regardless of the scale one uses to observe the system, on average it looks the same. But this turns out to define a mathematical problem of remarkable richness and complexity, e.g. [7]. This is the 'conformal field theory', the CFT in the AdS/CFT correspondence, as I advertised in the prologue. This is also a lively and modern research subject among pure mathematicians.

But this section is eventually supposed to explain *quantum* theory, so why worry about these classical physics constructions? The big deal is that nature is made from an *infinity* of things—in atomic physics and quantum chemistry deliberately circumstances are addressed where the most interesting work is done by the quantum mechanics of a few things. But how to get a handle on the quantum physics of such infinitely large systems? This kept physicists very busy in the second half of the twentieth century: this is the story of quantum *field* theory.

Quantum field theory (QFT) is typically the last course that is offered in the physics master programs—it is the last grand accomplishment of mankind in mobilizing equations to understand reality. It is the mathematical fundament of the standard model of high-energy physics with its quarks, Higgs bosons, neutrinos, vector bosons, and so forth that landed triumphantly in the 1970s, supported by the data generated by the enormous particle accelerators.

The understanding of QFT is still limited to special circumstances. We do not have mathematical tools that allow us to solve it in general. In the first place, little is known in general regarding non-equilibrium circumstances. The success portfolio rests on equilibrium experiments revealing typically *linear response* properties: the system is perturbed by infinitesimally small external forces that only trigger the response measured in the experimental machine. But since the system itself is not changed, because the perturbation is so small, this observable response can be computed relying entirely on the equilibrium theory.

The next problem is actually rooted in quantum statistics: the fermions versus bosons affair. As it turns out, when dealing with strongly interacting fermions at a finite density the mathematical problem is characterized by exponential complexity (the 'fermion sign problem'), with the ramification that, by principle, a quantum computer is needed to solve it. This has kept me quite busy during my career. Electrons in some solids, such as the high Tc superconductors, suffer manifestly from the sign problems, while experiments indicate that these have quite strange properties. The ADS/CFT ploy I discussed in the prologue, that strange mathematical machine invoking black holes as calculators, allows us to have a look behind the 'sign problem brick wall', being quite suggestive regarding the experimental 'strangeness'[7].

But this is work in progress. The triumphs of QFT are exclusively associated with the equilibrium physics of 'sign-free' systems. The reason is that this can be computed reliably to arrive at precision comparisons with experiment—among others, the standard model story. The reason for this is in the powers of Boltzmann's statistical machinery. This part of QFT is just a reinterpreted version of statistical physics, allowing the powerful stochastic equations to do the hard work.

This is not that well disseminated beyond the theory floors, the reason being that it relies on the *path integral* representation of quantum physics. Path integrals are viewed as fanciful and not all physicists take this course. This invention, that turned Richard Feynman into a celebrity, is however much liked and used by theorists, given that it is conceptually (not necessarily computationally) quite transparent, *because* of the intimate relation with statistical physics. In fact, this relationship was only realized in full in the 1970s. It revolutionized the understanding of quantum field theory in general, accelerating the phenomenal success of the standard model of high-energy physics in this era. I really got into it in the 1990s, inspired to quite a degree by the discussion of the quantum critical state in this language, by Sachdev [9].

How do path integrals work? These yield a rather intuitive, vivid view on the way that quantum physics generalizes the classical world (Fig. 3.4). Consider some system formed from elementary constituents, think particles. These form a world characterized by particular spatial configurations of these constituents that do change in time: the 'time slices'. Think about it as a movie, but now form a stack of these spatial images. For simplicity, focus on two-dimensional space like a film reel, forming in this way a cube, where time corresponds with the stacking direction of the various time slices. Such a 'cube' is called a world-history.

In classical physics a particular world-history is singled out, given particular initial conditions and the forces acting on the system: such a world-history is captured by the equations of motion of Newton's mechanics. But not so in quantum physics! In fact, this becomes most transparent dealing with equilibrium conditions. In classical physics one solves the equations of motion governing the motions of all microscopic particles, to discover that at long timescales this

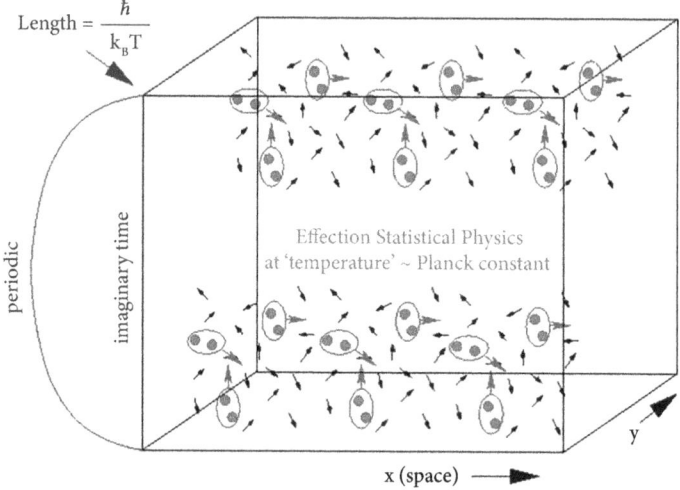

Fig. 3.4 Cartoon of the Euclidean path integral underlying thermal quantum field theory. This lives in space-time, where two space dimensions x, y are indicated. The extra dimension is the *imaginary* (or 'Euclidean') time τ, curled up in a circle with a radius set by inverse temperature. Inside this Euclidean space-time lives an effective statistical physics problem (in this cartoon formed from spins and particle pairs) at an effective temperature proportional to Planck's constant $\hbar$.

becomes chaotic and ergodic such that one can use Boltzmann's 'timeless' stochastic description. The quantum generalization is in a way remarkably simple. The essential generalization of quantum physics is in the fact that not one particular world-history is realized, but a-priori *all possible* world-histories may be realized!

This encapsulates the Schrödinger cat wisdom, a story that may be familiar to the reader: the animal that is dead and alive at the same time, until somebody takes the effort to have a look. But this now generalizes to all these different histories that can be supported by a thermodynamically large number of particles. The number of such histories is *exponentially large* in the already extremely large Avogadro's number (there are of the order of 10^{23} atoms per gram of matter). This is the origin of the hardship to compute QFT properties in general (like the non-equilibrium, sign problem), when we cannot profit from the simplifications of Boltzmann's stochastic machinery.

In the real time of Lorentzian signature, one then finds that all these world-histories behave like wave-like objects that tend to destructively interfere. The criterion for this is associated with the 'action'—discovered in the nineteenth century; this refers to a fanciful way to derive Newton's equations of motion. The quantity 'action', has the dimension of 'energy times time' and the equations of motion of Newtonian physics are recovered by demanding that it becomes as small as possible: the 'principle of minimal action'.

Now it comes: to decide whether the action is 'large' or 'small' in the quantum physics context one has to compare the value it takes for a particular world-history, with *Planck's constant*: the famous $\hbar$ often presented as the icon of quantum physics. In scientific notation it is as small as 1.1×10^{-34} Joules times seconds. You may have not much of an intuition what such numbers mean, but I can assure you this is extremely small as compared to the action associated with any macroscopic world-history. Under these circumstances, all world-histories, except for the minimal action classical ones, interfere away and all what remains is the classical universe! Isn't this stunningly informative? Our classical world is just a greatly handicapped version of the general quantum world, which in turn is characterized by much more freedom and richness. Anything that can happen will happen, be it at 'the same time' in the Schrödinger cat guise.

To compute matters in this original Feynmanian path integral representation is less straightforward—a reason that path integrals are still viewed as fanciful devices that appeal to the mathematically inclined ones. This is rooted in the hardship of keeping track of the destructive and constructive interferences of the 'world-history waves'. But one also encounters matters of principle that are not quite obvious from this formulation. Is this path integral computing how things work in equilibrium circumstances (including linear response), or can it cope with a causal time evolution? Given that it should know about equilibrium, how to do justice to the factor of finite temperature?

In the 1950s, in a gradual development involving a next set of unsung heroes of physics (Matsubara, Kubo, Wick,...), it was gradually realized how to generalise Feynman's path integral in order to cope with the difficulties alluded to in the previous paragraph. It was already precisely known how to factor in the 'quantum' in the finite temperature realms using the traditional 'canonical representation' of quantum physics. I have actually been deliberately avoiding this language here, for the simple reason that it is way more abstract-mathematical, involving non-commuting operators acting on the abstract Hilbert space. It was then found out how to reproduce these results using the path integral language. The outcome is stunningly simple and elegant [10]!

All one has to do is to Wick rotate the space-time of the path integral from *Lorentzian to Euclidean* signature, the affair that I already highlighted in my discussion of gravity (Section 2.5). To appreciate this in the path-integral context, one has to have some familiarity with complex numbers, but otherwise it is quite elementary. The Lorentzian time path integral is about sums where the terms are weighted by oscillating factors $\sim e^{iS(j)/\hbar}$, where $S(j)$ is the (Lorentzian) action of world-history j and $i = \sqrt{-1}$. Upon Wick rotating to imaginary time (Euclidean signature) these factors change into $\sim e^{-S_E(j)/\hbar}$ where $S_E(j)$ is the Euclidean action. But now you recognize the Boltzmann weights p_i in the classical partition sum! The sum over Euclidean world-histories turns out to be the quantum partition sum, the object one wishes to compute.

But you notice that $\hbar$ takes the role of temperature in the analogous classical problem. How does temperature enter in the quantum partition sum? The greatest surprise of all is that what remains to be done is to 'compactify' the imaginary time axis in a *circle with a radius* $R_\tau = \hbar/(k_B T)$: $\hbar$ has the dimension of energy times time, while $k_B T$ has the dimension of energy, and therefore R_τ is a time-like quantity. Dealing with a 'sign-free' system, the full quantum generalization of Boltzmann's partition sum is now given by an object that is a very close cousin of the classical partition sum. Instead of the potential energy of a spatial configuration E_j, determine instead the action as computed for the Euclidean time rolled up in the finite temperature circle, and call the outcome $S_E(j)$ for world-history j. The quantum partition sum becomes now $Z_\hbar = \sum_j \exp -(S_E(j)/\hbar)$. The take-home message is that it is in essence the same stochastic problem that has been so successfully coped with in statistical physics. But one has now also to deal with the 'springs' connecting the time slices of the world-history, while the probability for a particular configuration to occur is no longer associated with temperature and potential energy, but instead $\hbar$ and the Euclidean action of a world-history. Last but not least, temperature is encoded by rolling up the imaginary time dimension in the circle.

The take-home message is that equilibrium quantum physics, lacking a sign problem, is just statistical physics in disguise, to be computed in a 'space' with an extra time dimension that has a finite duration at finite temperature. In order to get at physical properties, one has to work through entertaining re-interpretations rooted in the Wick rotation, but we know how to do this (for highlights, see e.g. [9]). We also learn that the elementary path integral is actually only good for computing equilibrium physics, including linear response. In fact, much later it was found out how to compute quantum non-equilibrium with the so-called Keldysh–Schwinger path integrals, but these are way more laborious, and much less powerful given the inherent mathematical problems.

Let me stress once more that equilibrium quantum systems are only statistical physics in disguise when these are 'sign free'. The issue of the sign problem, which arises in fermion systems, is that individual probabilities in the quantum partition sum are no longer positive definite. For typical problems roughly half of the world-histories are characterized by negative probabilities and these of course do not make sense. The ramification is that the powers of the stochastic calculus underlying the success of 'stat phys' are no longer in effect and nearly nothing is known in general regarding sign-full quantum problems—my life-long obsession [7].

But when this mapping to statistical physics succeeds, mankind appears to understand literally everything that really matters regarding equilibrium physics departing from quantum physical principles. It is telling us how to relate the physics at finite temperature to the quantum realms at exactly the absolute zero of temperature, zero Kelvin or −273.15 degrees centigrade. Surely solids persist down to zero Kelvin but, after a long quest in the twentieth century, the 'conventional' quantum liquids were also deciphered: superfluids/superconductors and the Fermi-liquids associated with the zero-temperature metallic state [11]. Fermi-liquids are tricky—formed from sign-full fermions. However, the superconductors and closely related superfluids were figured out by the Landau school (Fig. 3.3) as being governed by an order parameter. It is just formed from the wave (instead of particle) side of the particle–wave duality of the microscopic bosonic particles. The order parameter theory also encodes the 'emergent rigidity' associated with order. The breaking of translational symmetry, characterizing solids, gives rise to *elastic* responses to shear forces, instead of the dissipative response of a liquid. Historically, the triumph of the order parameter has been in the elucidation by Ginzburg and Landau that the order parameter of the superconductor gives rise instead to a rigidity that renders currents to become dissipationless—the supercurrents that runs forever.

This becomes really impressive, dealing with systems that are very complicated. Eventually one needs computers and the powerhouse in dealing with stochastic problems is the so-called Metropolis Monte-Carlo algorithm, developed already in the 1940s as part of the Manhattan project aimed at constructing the atomic bomb. This flourished in the thermal/classical statistical physics context and it is called 'quantum Monte Carlo' (QMC) when it is used to compute the Euclidean path integral as a statistical physics problem in $d+1$ dimensions. This has produced outcomes that, more than anything else, demonstrate that mankind is capable of truly understanding the way that quantum physics works, of course allowing for the limitations of the no-sign problem and equilibrium.

One example that is less well known is the QMC results for Helium four. Helium is the only atomic system that is not solidified at zero Kelvin, and instead it forms a superfluid. The isotope with four nucleons is a boson and therefore the problem is sign free. However, this fluid is also dense and very strongly interacting: even the classical 'van der Waals fluid' is very tough to compute. But in the 1990s QMC was used triumphantly to compute all its equilibrium properties, matching quantitatively with experiment [12].

But, everybody will agree, the real triumph of Euclidean field theory is embodied by the QMC aimed at quantum chromo dynamics (QCD). QCD is part of the standard model, dealing with the quarks and gluons that form nuclear matter. It famously exhibits the *confinement* phenomenon: it takes an infinite amount of energy to separate a proton or neutron into its quark components. This is a physics special to the so-called non-Abelian Yang–Mills Gauge Theory: structuring the standard model and the confinement phenomenon was rewarded with several Nobel prizes.

It is yet again effectively sign free, and it can be formulated as a special kind of 'spin-like' model living on a four-dimensional lattice, which can be tackled by QMC. This spurred an intense computational research effort by a large community that triumphed not so long ago by the demonstration that even the difference in mass between the proton and the neutron rolls out of the computer [13]. These masses are actually rooted in quantum fluctuations of a most severe kind and it is quite an achievement that mankind can handle this, be it with the help of supercomputers. Yet again, I like to emphasize it because this is what I perceive as 'really understanding what quantum physics is', in so far we are capable of understanding it.

3.2 Unitary time evolution: the never ending quantum dance

I am done with the short history of mainstream quantum physics, as it culminated in cracking the QCD affair. But what has this to do with the main theme of this text, the meaning of time, especially in quantum physics? The take-home message of the above is in fact, that dealing with the 'timeless' equilibrium conditions, lacking causality as well as information processing, is what we are most comfortable with in quantum physics. We just understand it completely, the case in point being that we can reconstruct intricate quantum physical phenomena such as quark confinement in QCD, in an impressive way reconstructing the workings of nature.

Let me stress again that *time* is in the driver's seat. The splendour of the path-integral respresentation is in the fact that, as compared to the strictly static, 'high' temperature realms of classical statistical physics, there is still a time axis, and the quantum action unfolds in space-time. However, given equilibrium or alternatively the unitary time evolution, one finds out that one is better off with the Euclidean signature (imaginary) time. In a way this intrigues more than anything else: in Euclidean signature, causality disappears, as vividly illustrated in the GR context (Section 2.5), and by the eternal cycling around this time circle from the future to the past. The only observable remains is the *radius* of the time circle that encodes for finite temperature, $R_\tau = \hbar/(k_B T)$. Temperature is 'time-like'—I do not perceive this as completely self-evident. Surely energy and time go hand in hand in quantum physics—these are conjugate as expressed by e.g. Heisenberg's time-energy uncertainty relation. Temperature is like energy, but it is more than just energy. It goes hand in hand with the stochastic, dissipative attitudes associated with thermodynamics.

I like to refer to this time circle as the 'tranquil real quantum time', whatever. But it is only part of the story of time in quantum physics. Part II is about what happens when this pure quantum 'unitary universe' is forced to give in to the demands of the *causal* time of the big world. We have now to work much harder and this part is just lacking the grand discipline of either equilibrium quantum physics or GR. In fact, there is a habit of brushing these troubles under the rug: on the level of the fundaments, non-equilibrium quantum physics is quite improvised. A pragmatic rule book is available, that does work impeccably, but it is littered with mystery. One side of this unsettled affair is the 'collapse postulate' that I will discuss at great length in the next chapter. But there is another mystery—also during the unitary evolution, one is obliged to wire in causal structure in the quantum physics. But very different from e.g. GR, in all circumstances we have to stick this in literally *by hand*. Let me first add some more information regarding the mathematical machinery before delving into this affair in the next part.

Since path integral technology is less useful in these realms, let me lean instead on the canonical formalism. It is an abstract mathematical affair, detached from any form of human intuition. It took me as a student two years or so before my brain eventually accepted it, to then turn into a deep appreciation for its elegance. It is characterized by an amazing underlying simplicity. It departs from an infinite-dimensional Hilbert space, endowed with particular properties that the layman can safely ignore. The various axes spanning up these dimensions correspond with the same microstates as in stat. phys., enumerating *anything* static that can be realized, like the world-histories, but now lacking a time axis.

The state of the system as a whole is a big arrow pointing in some direction in this infinitely dimensional space. Now it comes: unitary time evolution just means that this arrow is in time just rotating in this infinitely large space, with a rate and a direction that is governed by the Hamiltonian. The Hamiltonian is in essence the total energy of the system, but now in a quantum operator incarnation, which means that it generates the 'rotations' when time proceeds. This maths is then insisting that the system is characterized by the quantized values of energy I already referred to. Notice that this formalism is strictly equivalent to the path integrals; one typically switches back and forth pending the problem one wants to address.

This unitary time evolution is sometimes called 'super deterministic': starting in a particular direction in Hilbert space at the initial time, it is precisely determined where the state vector is pointing at any later time. A more precise and modern way of addressing it is in terms of information theory. Shannon entropy is a key quantity in classical information theory and this is supplemented by the von Neumann entropy in quantum information $S = -\text{Tr}(\rho \log \rho)$, where ρ is the total density matrix of the system. It is easy to prove that the von Neumann entropy is *stationary under unitary time evolution, $dS/dt = 0$*. The constancy of this entropy implies that during the unitary evolution, information is *not* processed. This is a key ingredient of the quantum computer: the computation actually takes place at the 'read out', when the wavefunction is forced to collapse.

I find this quite fascinating. The capacity to process information goes hand in hand with causality: any computation is a causal affair, with an input at an earlier time, changing into an output, that is informationally different, at a later time. But unitary time evolution is devoid of causality. Without causality no information processing occurs, and when information is not processed, it is impossible to identify a causal order. Is it actually possible to distinguish sharply between 'computing' and 'causality'?

Once again, the big deal is that the quantum physics that keeps our quarks into nucleons, the electrons into atoms, the atoms into molecules, and so forth, is coincident with the unitary side. The electrons orbiting in molecules do so without any reference to a causal time, at least as long as external influences born in the causal macroscopic world do not interfere, causing chemical reactions. The quarks forming the protons and neutrons in your body stay hidden forever through their 'timeless confinement dance' as long as these are not pushed to extremely high energy in the LHC accelerator in Geneva.

But where are the probabilities of quantum physics, that are so much in the foreground in popular accounts? In fact, there is nothing of the stochastic kind going on in the 'unitary universe'. The reader may argue that the Monte Carlo algorithm can be used to compute the path integral that for zero temperature describes the ground state wavefunction; the unitary object *par excellence*. But this is in Euclidean signature; after the Wick rotation to Lorentzian signature, magically these QMC probabilities turn into 'coherent' wavefunction amplitudes, enumerating how the various states are represented in the overall wavefunction. There is surely stochastic wizardry going on with the finite temperature path integral associated with the imaginary time circle with finite radius. This describes a *thermal* state and this is not different from Boltzmann's classical thermal state in the sense that it hinges on the thermal probability factors p_n. But this sense of probability associated with equilibrium at finite temperature has, in the first instance, nothing to do with the 'God that is playing dice' paraphrasing Einstein's defeat by the Bohr school.

3.3 God is playing dice and the experimentalist is playing God

As Einstein complained to Bohr in his opposition to the Copenhagen interpretation, 'our Lord does not play dice'. In the way that quantum mechanics is taught, the inherent stochastic nature is much emphasized. But this is entirely associated with postulate number two of quantum physics: the collapse of the wavefunction as governed by the Born rules. The wavefunction is the unitary object, and these are only collapsing when a 'measurement is done'. But in order to execute a measurement, we need the extreme non-equilibrium circumstances that are present on our planet, eventually due to the finiteness of Newton's constant.

Surely the measurement problem is bigger than life—much of the text that follows is dedicated to this cause. But I am prejudiced that, for historical reasons, it is biased towards a particular context that may to a degree obscure its meaning for the workings of nature in general. It is standard lore among physicists: the various incarnations of the Schrödinger cat, the Stern–Gerlach experiment, the double slit experiment, all the way up to the loophole-free tests of Bell's inequalities. These all have in common that they have dealings with one (the cat), two (Alice and Bob), or three (Alice, Bob, Eve) qubit like degrees of freedom, if possible captured by some version of the two-level problem.

Once again, this tradition emerged in the 1930s, when only the quantum mechanics of small atomic-like systems were understood sufficiently well. Let me again emphasize that nature is inherently associated not with two or three, but instead by an infinite number of 'qubits'. The atomic physics-like experiments are in a way all very contrived, not at all representative for nature in its 'natural' condition. To execute these experiments, one needs a big lab space filled with intricate instruments, fine-tuned to reveal the Born rules at work in atomic realms. Their construction and operation, in turn, require copious amounts of low entropy photons from the Sun, which in turn are expressions of the 'gravity drive' keeping the causal evolution of the universe going.

What is a measurement? In the way this is handled in the context of the Born rules, there are intrinsic vagaries. One has to specify which quantities are measured, like the 'direction of the spin of the particle', but these have an intrinsic reference to the small, artificial atomic physics system. In a very concise way, what are the Born rules? These insist that one should define the directions of the Hilbert space to be associated with the quantum numbers that are measured (e.g., the 'direction and magnitude of the spin'). One should then take the amplitudes of the coherent superposition of the unitary wavefunction, and their squares then reveal the probability for a particular outcome of the experiment, like whether the spin is up or down.

It is uncontroversial that this 'collapse' has a satisfactory operational definition—the computed outcomes fit experiment—but at the same time it is still quite mysterious, begging for a bigger story explaining what is really going on. My impression is that this sense of mystery can count on a consensus in the physics community, and I will devote a whole chapter—the next one—to this multi-faceted problem.

Somehow, the collapse of the wavefunction is tied to the point where the acausal, timeless unitary world of 'pure' quantum physics is forced to relate to the macroscopic world with its causal time axis. This is the first deep question related to the difference in the role of time in the microscopic (unitary) realms and the gravity-driven time in the macroscopic universe. But there is a second incident where the tension between quantum physics and causal time is bigger than life. This seems to be not at all widely realized among physicists. It is as if we are brainwashed during our physics education to stare away from it. Besides the claim that gravity is responsible for causality, this is a second overlooked no-brainer that I have in the offering.

In the gravity chapter I highlighted, for a good reason, the amazing fact that the Einstein equations actually imply a universe that is intrinsically characterized by a causal evolution, a time that has therefore automatically a direction. It is however completely obscure how and why such a causality notion should arise as some kind of automatic necessity in quantum theory. The quantum theory is supposed to be a fundamental theory explaining the way that nature works. But macroscopic nature is characterized by causal evolution and what has quantum theory to say about the origin of causality?

The answer is, *nothing at all*! We can surely reconstruct the atomic physics style experiments in gory detail. But how does the quantum theorist cope with the fact that things are happening? The answer is embarrassing: he just sticks it in the equations *by hand*!

Time evolutions are governed by the Hamiltonian, but Hamiltonians in their natural incarnations just describe the 'timeless' unitary evolution. In order to track the causal evolution associated with the experiment, like the experimentalist switches on his laser and starts to count shortly thereafter the photons that enter his photodetector, one just invokes a 'parametric' time dependence in the Hamiltonian itself. This means that, say, the light field coming from the laser corresponds with a term in this Hamiltonian (it represents energy). One then just makes up a time associated with this causal evolution: when this time is smaller than zero (the 'past') the parameter in front of this laser field energy is adjusted by hand to disappear. At time zero it is switched on and its ramifications for the collapse of probabilities is then computed at the positive times.

Once again, this is the standard procedure that surely explains the outcomes of atomic physics experiments in gory detail. In fact, the quantum computer should be regarded as the twenty-first century incarnation of this affair: all the quantum gates are of this parametric kind, manipulating this aspect of causality in the unitary evolution. As a caveat, it is actually still the case that also in the presence of a parametric time dependence, no information is processed during the unitary evolution. A quantum computer computes at the 'readout', the instance where the wavefunction is made to collapse.

But, as a 'complete' theory of reality, quantum physics should explain the origin of this parametric time dependence. The standard reasoning, that it is there because the experimentalists is turning knobs, does not of course qualify as an explanation. But the situation is worse than even in the equilibrium classical realms—it is completely obscure. Unitarity itself excludes any form of causality, while the collapse process appears to generate randomness instead of the organized causal evolutions. Although it is not quite spelled out in the textbooks, in this regard quantum theory is painfully incomplete. This should be regarded as a central challenge for the theory that may well turn out to be quantum gravity: reconcile the causal nature of classical GR with all the aspects of nature explained by quantum physics, in a way that also the factor of causality becomes a natural part of the latter. I will not come back to this, for the reason that I have really nothing more to say about it—I have not the faintest clue.

3.4 The Hawking temperature of the black hole and the Euclidean time circle

The intention of this chapter has been to put the unitary time evolution as the central wheel of quantum physics in the limelight. The reader should now have a reasonable impression of the 'tranquillity' of this acausal part of nature. But there is not much of a connection with gravity yet. There is, to the best of my knowledge, one result that in a stunningly harmonious fashion ties together this unitary affair with a seemingly very causal gravity result. The outcome yields an utterly efficient mathematical explanation for perhaps the most famous 'quantum-gravity' result: Hawking radiation, and more generally, the claim that a black hole behaves like a thermodynamic entity endowed with a free energy that revolves around the Bekenstein–Hawking entropy.

What is this Hawking radiation? I will discuss the original derivation of this radiation by Hawking in more detail in the final Chapter 5. This follows the canonical route, involving assumptions regarding the way that quantum physics deals with the causality structure of the black hole in the sense of the Penrose diagrams (Section 2.4). In very short summary, one departs from the classical GR black hole geometry and inserts (free) quantum fields to then insist that the information contained in these fields behind the horizon are not available to an observer in the external geometry and can therefore be 'traced out'—actually a tricky affair as I will discuss in Chapter 5. It then follows that an observer at radial infinity from the black hole perceives this as a thermal state.

The black hole behaves as a thermal object, as a so-called black body radiator giving off radiation associated with a hot object, like hot metal becoming red. The temperature is the Hawking temperature associated with the surface gravity of the black hole. But thermal objects are characterized by a thermodynamical entropy and in a similar fashion one can address this entropy. This is consistent with the temperature: famously the entropy of the black hole is proportional to the *area* of the horizon. The entropy of a material body scales of course with the volume occupied by it. This 'missing dimension' is the birthplace of the idea called the 'holographic principle' behind the AdS/CFT correspondence that I will glorify in Section 5.1.

It is quite some work to compute these matters and, once again, I perceive it as resting on quite tricky assumptions. But there is another way of finding the same outcomes that is technically extremely simple and, at least in my perception, completely watertight. When you ask my opinion, this is the only quantum gravity-like result that I completely trust; I cannot discern anything suspicious in this derivation. This deals with the *eternal* (Schwarzschild) black hole.

It is so simple. The alert reader may already have wondered whether there is any relation between the time circle associated with this Schwarzschild geometry in Euclidean signature (Section 2.5, Fig. 2.6) and the time circle coding for finite temperature in the Euclidean path integral, Fig. 3.4. As I already stressed, the Einstein equations are equally valid in a geometry with Euclidean signature as in one with a Lorentzian signature. That the signature of nature is Lorentzian is just an empirical fact. But then we learned in the present chapter about the Euclidean path integral as a universal and convenient way to describe the quantum physics of matter in *equilibrium*. We learned that this includes the nature of finite temperature equilibrium, obtained by just rolling up the time axis into a circle with radius $\hbar/(k_B T)$.

Let us focus in again on the Schwarzschild black hole. I discussed at length in Section 2.4 the perplexing causality structure as given away by its Penrose diagram. But there is actually a tricky twist. This causality structure is the one associated with a 'probe', as it is called in the gravitational literature, actually closely associated with the linear response notion in quantum field theory (e.g. probes in the gravitational bulk are dual to the linear response function in the boundary of AdS/CFT). This is a good word: a probe is an object not carrying energy itself, and thereby it does not change 'by backreaction' the geometry as implied by the Einstein equations. In a strict sense these are unphysical objects that have only a meaning in terms of describing what happens when this limit is approached.

The Schwarzschild geometry, where there is only a black hole present, is therefore all there is. But this geometry is stationary, as a whole it does not change in time! Its mass, horizon radius, and so forth are, for all of eternity, the same. It is therefore associated with an impeccable *equilibrium state*! But we learned that in equilibrium, the Euclidean path integral will tell everything regarding the way that the quantum physics works. Let us therefore see what happens at radial infinity in a Schwarzschild geometry with Euclidean signature. Its imaginary time axis is rolled up in a circle with a radius set by the black hole mass. This means that the equilibrium quantum fields will be at a temperature set by this radius and sticking in the numbers at radial infinity one finds that this temperature is precisely the Hawking temperature!

In the culture of the (quantum) gravity community this is typically presented as a convenient technical short cut; instead of grinding through the canonical exercise, in a matter of seconds one identifies the Hawking temperature given the Euclidean Schwarzschild metric. But I find this as a bit of a deception in the context of the general role of the Hawking radiation in the quantum gravity theatre. When it comes to the role of time, I discern here a deep conceptual motive. Once again, the issue is that causality *does not exist* in Euclidean space-time. The time-like quantity that does persist is *temperature*, itself a 'timeless' quantity since it is only defined in an equilibrium state and in real equilibrium it is impossible to realize causal evolutions. This all fits together consistently dealing with the eternal black hole. However, when dealing with less tranquil circumstances, like when black holes do have a history and observers do back react, matters are less straightforward as I will discuss in Chapter 5.

4

The measurement postulate: God is playing dice when cause yields effect

As I explained in Chapter 3, quantum physics is a two-stage affair. The first stage is the unitary universe of 'pure' quantum physics which has been so greatly successful in explaining the nature of the substances in the universe under the 'timeless' equilibrium conditions – (Section 3.1). The second stage deals with non-equilibrium states, and what has to be done to connect this unitary universe to the macroscopic world, as characterized by causal evolutions and the processing of information. This requires separate ingredients. I expressed my dissatisfaction in Section 3.3 with the practice of wiring in parametric time dependences by hand in Hamiltonians. Again, as a preliminary condition since the time evolution continues to be acausal—no information processing—as long as it is unitary. This is then followed by the 'collapse of the wavefunction' involving a completely independent set of assumptions, captured by the Born rules.

I already complained that the 'measurement problem' has been suffering from a degree of tunnel vision, associated with handling the tiny quantum-mechanical special effect systems of atomic physics, where it all started. This is reflected in the Born rules: the way they are formulated gives away certain schematics, rooted in the simplifications of atomic physics. Citing the Wiki page [14]: 'The observable corresponds to a self adjoint operator A with eigenvalues and eigenstates $\lambda_I, |\lambda_i\rangle$. The probability for the outcomes of a measurement to yield λ_i is $\langle \lambda_i | \Psi \rangle$.' The trouble is with appointing the operator A, typically an object that is manufactured in the laboratory by an experimentalist, targeting physical properties like the direction of the spin (magnetic dipole) of a material particle, or the polarization direction of a photon.

I am myself more comfortable with the way that the 'collapse' is handled in the context of quantum field theory. This is just much closer to the natural state of reality. I did 'hard sell' its success in Section 3.1. One has to do experiments to find out what is going on and I already introduced the 'linear response' mode of experimentation as the quantum incarnation of the gravitational probe. The system is disturbed in an infinitesimally weak way, such that an observation can be done that will not disturb the equilibrium state of the system. It is just a fortunate circumstance that many laboratory experiments rather precisely do yield this information. A simple example is the resistance associated with electrical current in a metal. This is governed by Ohm's law, so the voltage is equal to the product of

On Time. Jan Zaanen, Oxford University Press. © Oxford University Press (2024).
DOI: 10.1093/9780198920793.003.0004

current and resistivity. Double the voltage and the current will double, while the resistivity stays the same, since the current is kept so small that the system is not changing, for instance by heating. The resistivity is the linear response quantity.

These responses are associated with macroscopic observables and are therefore 'after the collapse' entities. Mathematically, these are yet again 'probabilities for the outcomes' called expectation values, associated with natural operators in the field theory (Φ) measuring quantities like the density. To be precise, in the jargon these are called 'vacuum expectation values' (VEV) and their precise definition, including finite temperature, reads: $\langle \Phi \rangle = \mathrm{Tr}\,(\rho\Phi)$ where ρ is the total (thermal) density matrix.

The above recipe surely works impeccably, but since the very beginning the 'collapse' has been subject of much debate that rages until the present day. Much of this tends to be quite philosophical, lacking any form of empirical underpinning. This started with the Copenhagen interpretation, to a degree just paying tribute to the Born rules but invoking in an uneasy way the 'observer'. It is as if the consciousness of the experimentalist that makes him/her switch on the measurement rig, is crucial for the outcome of the experiment to settle in one or the other realization. This was actually taken much further by the Californian 'hippies who saved physics' in the 1970s, speculating that this implies intricate relations between (Buddhist-style) human consciousness and the essence of natural reality. When this was raging, I was myself in the grip of the conformism characteristic for adolescents, being very busy trying to be a good hippie. I have a vivid recollection of how I got in the grip of these mysticisms, reading the best selling bible of this affair, the 'Dancing Wu-Li masters' [15].

There is more romantic mystique discernible, perhaps brought to a climax by Everett's 'many-worlds interpretation'. Every possible outcome of a quantum measurement is claimed to correspond with a real world and all these outcomes define together a multiverse where a particular observer is following a particular path. Prepare a Schrödinger cat state, measure it to find that the animal is dead and you continue living in the 'cat is dead' universe. But you could have found the animal to be alive, and this 'cat is alive' world is also a real world, existing besides your personal 'dead cat' place.

These 'observer-centred' collapse philosophies have in common that they found their initial inspiration in simple Schrödinger's cat-type atomic physics ploys. However, let me stress again that such 'cat measurements' are in a way very contrived. These need a lab space filled with lasers, powerful fridges, fancy spectrometers, and so forth. This is *not* nature in its daily, meat-and-potatoes incarnation.

The way that things evolve in our daily world has much more to do with steam engines. Anything that is happening in our human world is eventually governed by heat currents driven by temperature gradients, that can be in part converted into the useful work that we learned as a species to control superbly, to enhance our physical comfort. But the time evolution of such macroscopic heat engines is intrinsically of a stochastic nature. It is actually a very modern idea, to link this form of *macroscopic stochasticity to collapses that happen spontaneously* without intervention by human experimentalists in the many-body quantum dynamics.

In this chapter I will develop this set of ideas. The first step amounts to a sketch of the reasons for this to be a modern affair. This is just because of the rather complicated technical nature of the maths dealing with non-equilibrium conditions in field theoretic systems. This is the modern subject of quantum thermalization (Section 4.1). This is now widely disseminated, but this is different from the idea to link steam engine stochastic dynamics to spontaneous collapses (Section 4.2). I will then zoom in on yet another idea that was not long ago quite controversial, becoming in the meantime more mainstream: the notion that the wavefunction collapse is rooted in the conflict between gravity and unitarity. This is the 'objective gravitational collapse idea' (Section 4.3) championed in the first place by Roger Penrose, a mind that you have already encountered a number of times.

Up to this point I will be just reviewing established sets of ideas by other parties. But I have then original material to offer. It is yet again a no-brainer, being familiar with the preceding material. *The notions of repeated spontaneous collapses leading to thermodynamics and the gravitational collapse of wavefunctions appear to complement each other in a natural fashion.* After presenting the case in Section 4.4, in the final section, Section 4.5, I will engage in a first attempt to chase down observable consequences of these ideas.

As an apology to the non-physics reader, this part of the story is a bit more equation dense for the reason that much of this material will be new also for the physics readership. I need a minimum of equations to render these affairs credible for this company. Again, as a non-physicist you may just skip the passages with equations: I hope you can continue to track in this way the big picture.

4.1 The collapse and the many-body nature of reality: quantum thermalization

In hindsight it is a bit embarrassing. To my usually well-informed brain it is a rather recent realization that all of the business of heat is itself a ramification of collapsing wavefunctions. This is the notion of *quantum thermalization*, perhaps best captured by the notion of the Eigenstate Thermalization Hypothesis (ETH) as formulated by Deutsch and Srednicki in the early 1990s [16]. The reason that I did not learn it as a young student is just associated with the fact that the mathematical difficulties did shroud wisdoms that are in hindsight obvious.

The 'classic' (1930s) musings regarding the collapse revolved entirely around the simplest quantum-mechanical systems. Much of it (Schrödinger's cat, Bell pairs, ...) rest on the simple 'two-level problem' that can be solved on the back of an envelope. However, reality is not about two states but instead about roughly 10^{23} qubits (1 qubit is a two-level problem) per gram of matter, spanning up a many-body Hilbert space that is exponentially large in Avogadro's number: $\sim 2^{10^{23}}$.

Once again, the Hilbert space is the theatre where the unitary time evolution unfolds. Why is it exponentially large ($2^{\#}$) in an extremely large number ($\# = 10^{23}$)? Although a bit of a caricature, it is just fine for counting purposes to assume microscopic degrees of freedom to acquire two values, like the $0, 1$ of the qu(antum)-bit. Dealing with two such objects one can construct four possible 'worlds': $(0, 0), (0, 1), (1, 0), (1, 1)$. The dimension of this Hilbert space is $2 \times 2 = 2^2$. As you can count out for yourself, dealing with three objects one finds a total of eight such 'worlds' of the kind $(0, 0, 0), (0, 0, 1), \cdots$, corresponding with $2 \times 2 \times 2 = 2^3$.

In other words, for N qubits there are a total of 2^N 'worlds'. Given that there are of the order of $N = 10^{23}$ qubits (atoms and so forth) per gram of matter one arrives at this gargantuan dimension of the Hilbert space. To give you some feeling for this number, there are of the order of 2^{80} protons and neutrons in the universe. When the universe would be classical we would have to cope with solving 'only' 2^{80} Newtonian equations of motion, a task tiny as compared with finding your way in the enormous Hilbert space.

In fact, there is a beautiful mathematical discipline enumerating with high precision how difficult it is to solve particular problems: 'complexity theory'. In general, the quantum many-body problem belongs to the worst case class, the so-called 'NP-hard'. But you do not have to understand its subtleties: in essence it is just the fact that one has to keep track of an exponentially large amount of numbers that renders it impossible to be solved by a *classical* computer, let alone by a human scribbling symbols on the back of an envelope. This is part of the motivation for the intense engineering effort that is unfolding at the moment, to build a quantum computer. Such a machine is in principle capable of tackling such problems in a reasonable 'polynomial' time, referring to the effort scaling like a polynomial $\sim N^{\#}$ instead of an exponential $\sim \#^N$.

You may now be puzzled: in Section 3.1 I was hard selling the case that the great success of modern physics is in the detailed quantum field theoretical way of understanding how macroscopic reality is born from the microscopic constituents. Apparently we are capable of understanding certain aspects of the quantum many-body problem, resting on maths that work. But this is due to simplifications intrinsic to the underlying stat. phys. problem. The bottom line is that when dealing with stable phases of matter, invariably an enormous 'thinning' of the Hilbert space, relevant to the state of matter, is taking place. The Metropolis Monte Carlo computer algorithm is tailored to cope with such situations with near infinite accuracy.

This can be understood intuitively. Consider for example the crystal as a typical ground state of matter—imagine it at the absolute zero of temperature. In a crystal the atoms are at fixed positions, forming a regular lattice: this corresponds with just a single 'direction in Hilbert space'. The perfectly ordered state is therefore extremely non-complex, but in reality one has to cope with 'fluctuations around the ordered state'. A single atom can leave its crystal position for a short instance, to quickly get back in the registry. But such stochastic fluctuations are of polynomial complexity and Monte Carlo can handle this. It became the gold standard in classical statistical physics and likewise 'Quantum Monte Carlo' (QMC) became a powerhouse for the quantum problems.

Yet again, this simplification is special to the *equilibrium* physics of *stable* phases of *sign-free* quantum problems with their stat. phys. attitude. In other cases, the exponential complexity raises its head immediately. Even the critical state associated with stat. phys. phase transitions tends to be of this kind. But the real big deal is that typical *excited* states are invariably characterized by exponential complexity. This is the reason that we know so little about quantum non-equilibrium. Because of the exponential complexity we do not have mathematical tools that work universally.

Since it is impossible to solve anything rigorously, dealing with this exponential complexity, even a basic issue like the ETH is for this reason still a hypothesis. But it is a good one that is generally believed to be true. It amounts to the following statement. Depart from a typical interacting many-body system that is characterized by only a few globally conserved quantities. Prepare at time $t = 0$ a pure state that is a superposition of highly excited energy eigenstates, narrowly bunched together in a small energy interval. Subject this state to a unitary time evolution, to then determine the VEV (vacuum expectation value) of an arbitrary operator. The hypothesis reads: when one waits long enough *this VEV will be identical to that of the expectation value of this operator in a thermal equilibrium state*, at a temperature set by the initial temperature adding the temperature rise associated with the energy that has been invested.

This is very interesting. Applying the standard postulates of quantum physics in the form of a unitary time evolution and observables corresponding with VEVs by driving a typical quantum many-body system out of equilibrium, one finds that eventually the outcome is a heating of the system, a rise of temperature! This is in stark contrast with the conventional understanding of the origin of heat as based on classical microscopic physics.

The understanding of heat in classical systems revolves around the notion of classical chaos. This is familiar to all of us in the form of the fact that it is impossible to predict the weather over a period longer than roughly a week. This is rooted in the notion of the 'Lyapunov exponent', that an infinitesimal uncertainty in the initial conditions of the classical many-body equations of motions blows up in an exponential fashion as a function of time, causing the system to become completely unpredictable. But this also is at the origin of the 'ergodic theorem' that at long times such a dynamical system will be able to explore the whole of the classical configuration space such that the time averages can be replaced by the spatial averages that define the Boltzmannian partition sum. The second law of thermodynamics, insisting that entropy always increases, is then explained by a famous metaphor stressing the role of probability. An egg falling from a table shatters on the floor. What is the probability that this egg reconfigures itself in an intact form on the table? That is of course extremely minute since the intact egg is a much less probable state than the shattered egg.

Classically, the stochastic nature of thermodynamics is eventually a pragmatic affair. Since it is impossible to keep track of the equations of motion while these allow at the end of the day that anything that can happen will happen, we can get away with a statistical description. But now we turn to the quantum case encapsulated by ETH, discerning a striking contrast. Now the probabilistic nature of heat originates in the collapse of the wavefunction which in turn postulates that stochastic behaviour is *fundamental*: it is the God playing dice affair, referring to a famous quote by Einstein in his shouting match with Bohr!

In fact, the effective analogy between classical and quantum physics can be pushed a bit further. One can identify a 'quantum Lyapunov time' in the following way. Upon pushing the quantum system locally out of equilibrium it will take a certain time before the system starts to explore the exponentially large Hilbert space. The maximum amount of information that an observer can retrieve is after the collapse. These averages will contain an amount of information that is infinitesimal as compared to what is required to reconstruct the unitary evolution in the gargantuan Hilbert space. This defines the onset of quantum chaos since the ETH is insisting that the system behaves in this quantum chaos regime as a thermal system. The time that is required for this to happen is the quantum Lyapunov time.

In fact, to a degree the ETH development that started in the 1990s was preceded by the accomplishments of the 1950s dealing with the conventional quantum liquids mainly developed by the Soviet school of Landau [11]: the Fermi-liquid and the superconductor/superfluid. They addressed the macroscopic properties of the fluids at a finite temperature using the same ingredients of unitary time evolution and the VEVs. But they were exploiting the fact that these are very gaseous systems where one can use the quantum generalization of Boltzmann kinetic theory. The classical version is very easy and is typically taught in the undergraduate freshmen year. Gas physics is just revolving around the fact that most of the time the particles are flying freely and independently, while the typical time it takes for a collision, sets all the properties of the macroscopic fluid. For this reason the Landau school was successful despite their ignorance regarding the quantum thermalization principles.

In the way that the ETH is usually formulated it still invokes the observer (experimentalist) as an active agent in the sense that in order to identify the production of heat, an observer is required. But is this necessary? In fact there is nothing that insists on the presence of an observer. Could it be that these 'collapses raising the temperature' happen all the time spontaneously, not just when humans are turning knobs on machines? When the experimentalist takes the effort through the ETH experiment he/she observes the birth of heat in the form of the stochastic dynamics of the thermal system. But would it be the case that the gas clouds on the other side of the universe suddenly start to behave as gas clouds when these are first observed by a human looking at a computer screen showing the signals from the Webb space telescope? This sounds quite unlikely.

What we know for sure is that gas clouds are governed by the stochastic equations of irreversible thermodynamics. Given that this should also be the case when observers are *not* observing these clouds directly, shouldn't it then be the case that the gas cloud wave functions are all the time *spontaneously* collapsing in order to produce the heat?

4.2 Spontaneous collapses and the origin of thermodynamics

In fact, very recently a gang of theorists worked out this idea of spontaneous 'ETH-style' collapses being the origin of thermodynamics. These ideas appeared in a concise fashion in a review [17] that I will indicate with the name of the first author, D'Alessio *et al.* This review is quite popular, it seems mainly for the excellent discussion of the state of the art of the ETH as I just discussed in the first section of this review. However, it seems that the ground breaking second part, revolving around the relations between 'ETH collapses' and thermodynamics, has been barely disseminated in the community. I discovered it supervising a master student reading class, aimed at the ETH. I found it a revelation!

It may therefore well be news also for the professional physicist. For them to get it sharply in focus, some equations are needed, and I apologize to the non-physicists: you can actually safely skip these passages and continue reading when the equations have disappeared again.

The crucial work is done in Section V of Ref. [17]—the authors appear to be a bit shy, actually camouflaging their central assumption of the repeated spontaneous collapses in an innocent-looking subordinate clause. It is a highly interesting piece of backward engineering. They depart from the time-tested, highly precise stochastic equations of motions governing irreversible thermodynamics. There is a hierarchy, departing from the Fokker–Planck equation, governing Brownian motion, being in turn rooted in the (Chapman–Kolmogorov) master equation, all revolving around random forces and noises. But D'Alessio *et al.* stress that this is in turn derived from more fundamental equations, describing *double stochastic dynamics*. Hence, this dynamics underlies all of thermodynamics and one can now ask the question, what is actually its origin?

At this instance they identify the mathematical connection between this dissipative stochastic evolution of thermal matter and quantum physics. The double stochastic dynamics requires a 'microscopic' time step, where the 'dice are thrown', determining what happens at the next time interval. It is unclear what the duration is of this time step—it has to be very short compared to macroscopic times, but long compared to the timescales of the unitary evolution of atomic physics. They then postulate that this time step is associated with the *spontaneous* collapse of the many-body wavefunction.

It is just happening automatically without the intervention of an observer, whatever. After the collapse, the system settles back in a unitary time evolution influenced by a parametrical time dependence in the Hamiltonian, to then collapse again at the next time step associated with the double stochastic dynamics. Hence, the claim is that the stochastic dynamics of thermal matter is fundamentally of a probabilistic nature, originating in a progression of spontaneous wavefunction collapses.

These require a typical, yet unknown scale: at times short compared to this collapse time/duration of the stochastic time step, the dynamics has to be unitary since we just know that this is the way things work on the truly microscopic scales of atomic physics. At the same time, it has to last long enough that the unitary time evolution has entered the quantum chaotic regime. The time step has to be longer than the quantum Lyapunov time, which is required for mapping onto the double stochastic dynamics. For this reason the ETH principle plays a critical role as well in getting this mechanism in working order!

One more conclusion can be drawn directly from this construction. Since the macroscopic stochastic dynamics produces *heat*, it follows immediately that the measurement basis that has to be specified for the collapse is actually the basis of *energy eigenstates*. This is also appealing and elegant. There cannot be an experimentalist selecting such a basis by turning knobs on his/her rig, since these collapses are *spontaneous* and therefore somehow 'objective'. The energy basis has a unique status as being in a conjugate relation with time, and it feels right to assign to it this exclusive status.

Let me now sketch how this really works by sprinkling a couple of equations—non-physicists, start reading again when the equations have petered out. The key quantity is the total density matrix of the system. Let us depart at time $t = 0$ from a thermal (mixed) state with a density matrix with elements,

$$\rho_{nm}^{(0)} = \delta_{mn} p_n^{(0)} |n\rangle \langle n|$$
$$p_n^{(0)} = e^{-\beta E_n} \tag{4.1}$$

Switch on, at positive times, a term in the Hamiltonian that is parametrically dependent on time, $\hat{H}(t)$. A unitary evolution means that at some later time the density matrix becomes,

$$\rho_{\tilde{m}\tilde{m}'} = \sum_n U_{\tilde{m}n} \rho_{nn}^{(0)} U_{n\tilde{m}'}^{\dagger}$$

$$U_{n\tilde{m}} = \langle n | \hat{U} | \tilde{m} \rangle$$

$$\hat{U} = T_t \exp\left[-i \int_0^t dt' \hat{H}(t') \right] \tag{4.2}$$

T_t refers to time ordering, while the tildes refer to the eigenstates of the Hamiltonian after the time evolution. Obviously, departing from the diagonal thermal state, after the time evolution, the density matrix will become littered with off-diagonal matrix elements in terms of the energy basis, after the evolution. But let us now terminate this unitary evolution by assuming that a collapse occurs. Allow the system the time to 'run into the quantum chaos', such that one can invoke the dephasing according to the ETH prescription. Notice that in this stage we have identified the quantum Lyapunov time as the minimal duration for the collapse to occur, and we have identified the energy basis as measurement basis.

The key to ETH is the claim that after this dynamical evolution, followed by collapse and dephasing, the new density matrix $\rho^{(1)}$ is diagonal again in the new energy basis $|\tilde{m}\rangle$,

$$\rho_{\tilde{m}\tilde{m}}^{(1)} = \sum_n p_{n \to \tilde{m}} \rho_{nn}^{(0)}$$

$$p_{n \to \tilde{m}} = U_{\tilde{m}n} U_{n\tilde{m}}^{\dagger} = |U_{\tilde{m}n}|^2 \tag{4.3}$$

The ETH claim is that, if one waits long enough, the diagonal (mixed state) density matrix $\rho_{\tilde{m}\tilde{m}}^{(1)}$ will coincide with a thermal state characterized by a temperature, raised by the amount of absorbed energy. But that is right now of no further concern; what matters is that the combination of collapse and dephasing turns the unitary evolution Eq. (4.2) into a stochastic affair associated with a thermal-like mixed state.

The big deal is that the 'doubly stochastic' $p_{n \to m}$ matrices in Eq. (4.3) can be directly interpreted as defining a discretized form of the Master equation that governs a stochastic evolution of the correct form to describe the thermodynamical evolution of a classical thermal system. Discretized means here that the time derivative of the usual, continuous time Master equation is now discrete, involving the finite time interval after which the collapse occurs. The full time evolution is associated with *repeated* collapses, like

$$\rho_{nn}^{(0)} \xrightarrow{U_1} \rho_{mm}^{(1)} \xrightarrow{U_2} \rho_{ll}^{(2)} \cdots \xrightarrow{U_N} \rho_{kk}^{(N)} \tag{4.4}$$

The doubly stochastic matrices form a group; for instance, this sequence can be represented by a single doubly stochastic matrix s, obtained as the matrix product of the fundamental doubly stochastic matrices,

$$\rho_{kk}^{(N)} = \sum_n \rho_{nn}^{(0)} s_{n \to k}, \quad \mathbf{s} = \mathbf{p}^{(N)} \cdots \mathbf{p}^{(2)} \mathbf{p}^{(1)} \tag{4.5}$$

Once again, this suffices to describe the time evolution of thermodynamical systems and I refer to D'Alessio *et al.*, who go to great length to fortify this point. For instance, it is easy to find out that as long as work is exerted by the time-dependent Hamiltonian, the system will end up in the high temperature limit: entropy is maximal and this means that the second law is fully obeyed.

It should be by now clear why the nature of the 'measurement basis' for the projections is unambiguous. This basis has to be the *energy basis*. This is in fact also stressed by D'Alessio *et al.* and the reason is obvious: the stochasticity generated by the repeating collapses generates *heat* and this means that the diagonal density matrices $\rho_{ll}^{(j)}$ in the sequence Eq. (4.4) are *thermal* density matrices, and for this to be the case, the 'measurement basis' has to coincide with the energy eigenstates. To my strong opinion this makes full sense. In a lab, the experimentalist can manipulate a particular basis to become the preferred one. But nature has to accomplish it spontaneously, and so much is clear that *time* is the culprit. Energy is conjugate to time and thereby the energy basis arises as the objective basis for the projection. To call this a 'measurement basis' is just odd—the reason that I put in the quotation marks. It just reflects a persisting confusion, revolving around a perception that some human intervention is necessary that was born in the early days of quantum mechanics.

D'Alessio *et al.* deliberately shy away from intervening in any way with the 'foundations' debate, they even do not refer to it. But placing their no-nonsense line of arguments in this context, the ramifications are, at least in my head, staggering. One does not need a lab with fancy lasers and other devices to kick around isolated microscopic quantum systems to force them to become part of macroscopic reality, with the consequence that their wavefunctions collapse. Instead, the claim is that we sense it all the time in mundane everyday reality, for instance when we warm our limbs on a cold day at a heater!

4.3 Objective wavefunction collapse due to gravity: the Diosi–Penrose model

The main theme of this text is on the nature of time, and I hope that in the meantime I have convinced the reader of the very different way that the equations of general relativity on the one hand, and quantum physics on the other hand, deal with this factor. The strong contrast is associated with causality: the way it is hard-wired in the Lorentzian signature geometries and the Einstein equations of gravity, versus the 'timeless' (in the sense of non-causal) time, intrinsic to quantum physics, where causality and information processing have to be inserted by hand.

We have arrived at a point where we can close this circle further. In fact, the story that now follows has been to a degree a motivation for the narrative that has unfolded up to now. The basic claim is that general relativity and unitary time evolution actually *mutually exclude each other, on the level of the most basic principles forming the fundaments of both theories!*

My awareness of this conflict started when Roger Penrose, the mathematician that you already met in the context of GR causality structure, visited Leiden to present a colloquium early this century. He hammered home a story, which was back then quite controversial [18]. Resting on a Schrödinger cat-type model he explained that unitary evolution and GR geometry are actually mutually exclusive: these two theories cannot be complete and correct at the same time.

He then argued that eventually gravity would win. Dealing with the microscopic physics where unitarity rules, he argued that the factor 'mass' (or energy, it is all the same in GR) is so small that unitarity can survive. However, when objects become bigger, their mass is increasing and at some point the conflict with gravity kicks in. Given that nobody has ever observed unitary time evolution involving macroscopic bodies, he concluded that there should be a scale in between the microscopic and macroscopic dimensions where unitarity is 'defeated' by gravity and the wavefunction collapses. This is the notion of the 'objective' (no human observers required) gravitational wavefunction collapse.

He continued with a dimensional analysis, resting on a particular model [18], in the same era also introduced independently by Diosi [19]. This suggests a gravitational wavefunction collapse should occur in a time of the order of one second, dealing with a Schrödinger cat state formed from an object with the mass of an *E. coli* bacterium being in superposition with itself displaced over a distance of one micrometre. In the remainder I will call this the '*E. coli* collapse scale'. Anything appreciably smaller than this will stay in the eternal coherent superpositions of unitary time evolutions, but anything bigger than this will directly collapse. The beauty is that Penrose turned the wavefunction collapse and, on a deeper level the GR-quantum tension, into an affair that can only be addressed empirically. Thinking over these dimensions one discovers in no time that it is tremendously difficult to track such a 'mesoscopic' system in the laboratory. There are no data of any kind available that shine light on this matter.

This in fact triggered an experimental effort aimed at realizing such cats in the laboratory. Although way ahead of its time when it was first proposed by Penrose and Diosi in the 1980s, it has developed in the meantime into a mainstream affair where several labs are patiently investing, aiming to eventually reach this regime, mobilizing state of the art nano-technological means.

Where then is this fundamental conflict, hard-wired in the equations of general relativity versus quantum theory? In fact, in Section 3.4 I highlighted the case of the eternal Schwarzschild black hole that appeared to be in striking harmony with the principles of thermal quantum field theory, culminating in the Hawking temperature at spatial infinity. What can go wrong? In fact, this appears to be a singular case where the great troubles that are luring around the corner are effectively brushed under the rug. Likely the reader already anticipates the first speciality of this case: as I stressed in Section 3.4, the eternal black hole is a stationary GR solution. It can therefore be married with equilibrium field theory, avoiding the causal time that should otherwise cause troubles on the quantum side.

However, there is also another source of trouble. In the GR literature this is referred to as 'back-reaction'. One typically invokes the help of a 'probe' in GR to interrogate solutions, like the extended Schwarzschild geometry with its intriguing Penrose diagram as discussed in Section 2.4. But these are in fact idealizations. It is just impossible to create an entity that is capable of observing anything that has precisely a *vanishing* gravitational mass. A physical observer has always a finite mass and by just being present they will actually change the geometry: 'the observer will back react'. But now, you may argue, I am interested in a black hole with a mass of a couple of Suns and I can build a powerful satellite probing this system that weights only a couple of hundred kilograms. Why should I worry? Believing Penrose, the issue is that even such a small mass will have a singular effect on the quantum physics of the object, as long as it exceeds the *E. coli* collapse scale.

The issue is that upon taking into account this back-reaction, one runs into a devastating problem of principle. In fact: *general relativity and unitary time evolution mutually exclude each other.* This has been much emphasized by Penrose although it has until the present day been largely ignored by the quantum gravity mainstream. It is however a statement that follows directly from the mathematics and is therefore serious by default. The only open question is: *under which conditions does this conflict become manifest,* heralding the end of either GR, or quantum theory, or of both theories.

This departs from yet another counter-intuitive fact, associated with curved Riemannian manifolds endowed with a Lorentzian signature. One way of formulating it is as follows: 'No point-wise identifications are possible between geometries determined by different mass distributions'. Yet again, the Lorentzian signature of time is the rogue player. The essence is: 'The clock of an observer in a universe with a particular mass distribution will tick differently when this mass distribution is changed'. In GR one cannot identify an objective 'global' clock ('time-like Killing vector') that ticks the same way, regardless of the way that one packs matter and energy in this universe. However, in order to mathematically define a quantum physical unitary time evolution, there should be a *unique* global time regardless of the way that mass is distributed! This is the fundamental problem of principle.

The coherent superpositions associated with the unitary evolution will typically involve states representing different distributions of mass. If so, their disagreement regarding the way that their clocks are ticking differently will accumulate in the course of time, such that one cannot maintain the unitary 'coherence'. Of course, since we do not know the overarching theory of quantum gravity—necessarily free of this pathology—one can only speculate what happens next. Since gravity works impeccably on macroscopic scales, especially with regard to endowing this reality with causal structure, it is reasonable to assert that gravity wins. When this 'ambiguity of time' becomes noticeable, it may well mean that unitarity has to come to an end. This may in turn have the consequence that the wavefunction collapses.

Given the absence of a theory, the only way to make progress is by experiment. We know something here. Apparently, on the microscopic scale of atoms and molecules, matter stays coherent forever. But the mass scales are accordingly very small and Penrose's time ambiguity should tend to vanish. On the other hand, on the human scale, the factor of mass is apparently dominating to such a degree that unitarity is completely destroyed. Apparently, the regime where one could directly observe the wavefunction collapsing by the 'pull of gravity' is somewhere in the middle. As I already stressed, there is a huge regime 'in the middle' where it is for technological reasons extremely difficult to unambiguously observe such a gravitational wavefunction collapse.

One would like to quantify it, at least to the degree that one can arrive at a dimensional analysis indicating what has to be accomplished in the lab. The estimation of the 'time ambiguity' magnitude is not straightforward. Penrose and Diosi [18; 19] arrived at an estimate that is a bit hand-waving and surely far from being generally applicable. This specializes to a rather literal Schrödinger cat circumstance. Consider a body with mass M, with its centre of mass localized at position $\vec{x}$, in coherent superposition with itself being displaced to a new position $\vec{x}+\vec{l}$. Quantum physics is associated with $\hbar$ carrying the dimension of action, energy times time. Upon identifying a gravitational quantity with the dimension of energy that does relate to the ambiguity of time, by balancing it with $\hbar$ one will therefore obtain a timescale that may be identified with the gravitational collapse time. Penrose suggested using for this the gravitational self energy: keep the gravitational potential well, associated with the cat, at $\vec{x}$ frozen, and move its centre of mass in this potential well to the position of its quantum copy, $\vec{x}+\vec{l}$. The bottom line is that upon inserting the mass of E. coli, one finds that it will take of the order of a second for this state to collapse, when l is of the order of a micrometre. This is right in the middle of the 'forbidden' regime, posing maximal hardship for the experimentalists!

There is actually a more appealing way to estimate this Penrose collapse time. This ploy was figured out by Tjerk Oosterkamp, an experimental colleague in my Leiden physics department, and a dear friend of mine, who got infected a long while ago with the Penrose challenge. I helped him out perfecting the case, earning a co-authorship of the paper. It is actually the only paper on my literature list having only an arXiv coordinate [20], the physics preprint server. After being bounced by a variety of journals during a campaign that lasted a number of years, Tjerk eventually gave up his attempts to get this published. Frankly, I have never encountered such trashy referee reports in my career. Somehow, the community that is investing in these matters appears to be formed by rather obsessive characters in the grip of defending their own particular views with a religious zeal. The bottom line is that the story has been completely ignored: since 2013 it has only been cited five times, including three self-citations. Let me do here some hard selling; it is worth it!

This is a second subject in this chapter that deserves equations—once again, if you have a maths allergy, start reading again when the equations are over.

Tjerk's point of departure is a famous example of the time ambiguity as found in any GR textbook: Shapiro's time delay experiment, executed in the 1960s, representing the first direct observation of the ambiguity of time in GR. It just involves the bouncing of a radar beam from planet Venus. That takes a certain time, set by the Earth–Venus distance and the velocity of light. This time is easy to measure with a very high precision. Shapiro measured this time lapse in two constellations, one where both planets are on one side of the Sun and another one where they are on opposite sides of the Sun. In the second case the radar beam travels through the gravitational field of the Sun while in the first case, the Sun is far away. According to GR, the Sun's gravity will cause the clock to tick slower and the radar beam will take more time when the Sun is in the middle. This gravitational time lapse as measured by Shapiro, turned out to be precisely consistent with the GR prediction and were confirmed to even higher precision with its modern version (Fig. 4.1).

Tjerk realized that this Shapiro way of measuring time affected by gravity can be combined with a Schrödinger cat ploy. The appeal of his construction is that, different from the rather ad hoc gravitational energy of Penrose–Diosi, it directly addresses the time ambiguity associated with the different mass distributions. Tjerk's Gedanken experiment is as follows. Consider a ball of solid matter that can be in a large- and small-volume phase. Such solids actually do exist, an infamous example being plutonium. As it turns out, related to the peculiarities of its 5f electrons, it exists in a large- and small-volume phase, having a difference of 30% or so in density. At ambient conditions it is in the large-volume phase, but the small-volume phase is very nearby. Upon applying a small pressure of the kind characteristic for metal milling equipment, it suddenly shrinks spontaneously. You may guess it—in early 1945 or so the first plutonium 'pit' became available for the Nagasaki 'Fat Man' plutonium nuke. Imagine, this sphere had to be fine polished and suddenly it shrank below the critical volume, turning into a red hot highly radioactive monster ball. It is still classified what exactly happened...

Fig. 4.1 Cartoon of a contemporary version of the Shapiro time delay measurements. This now involves the Cassini satellite that is in the first place studying the Saturn system. But as a byproduct it can deliver high-precision data on the gravitational time delay—the figure speaks for itself.

For the sake of the thought experiment, we take such stuff and we manage to create it in a coherent superposition of the large- and small-volume phases. We surround it by mirrors and let light rays bounce back and forth in this cavity, using these in the Shapiro guise as clocks. Let us write down the coherent superposition describing this cat state—a and b represent the large and small ball, respectively:

$$|\psi(t)\rangle = \alpha e^{-\frac{i}{\hbar}E_a t_a}|a\rangle + \beta e^{-\frac{i}{\hbar}E_b t_b}|b\rangle \tag{4.6}$$

This is business as usual, with the energies of the two balls being different a priori ($E_{a,b}$), prepared in a way that the amplitudes α, β are different. The only novelty is the occurrence of *different* times for the large and small balls: $t_a \neq t_b$. This difference is of course set by the Shapiro time delays. A straightforward computation yields,

$$t_a - t_b \simeq t\frac{8GM}{2Lc^2}\frac{b}{a} \tag{4.7}$$

where G and c are Newton's constant and the velocity of light, respectively; a, b are the radii of the large and small volume ball with mass M while $L \simeq a, b$ is

Fig. 4.2 Tjerk Oosterkamp showing off some of his machinery that may eventually land the Penrose-style gravitational wavefunction collapse inside the observational window. Tjerk proudly considers himself in the first place as an engineer researching revolutionary new machines.

the distance between the mirrors. Dealing with unitary time evolution, physical observables only depend on the energy difference $E_a - E_b$. However, according to Eq. (4.6) the crucial difference is in the fact that the phase becomes sensitive to the *absolute* energies. Write $E_a = E_0$ and $E_b = E_0 + \Delta E$, assuming that $E_0 \gg \Delta E$. It follows that a non-unitary phase difference evolves according to,

$$\phi^*(t) \simeq \frac{1}{\hbar} E_0 \, (t_a - t_b) \tag{4.8}$$

What to take for E_0? In gravity the *absolute* energy matters and dealing with the ball with the large rest mass, $E_0 = Mc^2$. Together with Eq. (4.7),

$$\phi^*(t) \simeq \frac{t}{\hbar} 8GM^2 \left(\frac{b-a}{2La} \right) \tag{4.9}$$

The time τ_G where this phase difference becomes $\simeq 2\pi$ should signal the 'end of unitarity', the collapse time:

$$\tau_G \simeq \frac{4\pi\hbar La}{8GM^2(b-a)} \tag{4.10}$$

This scale turns out to coincide with the prediction by Penrose, based on his gravitational self-energy estimate applied to the present set up. This is apparently the hidden meaning of the gravitational self-energy in this context. We completely failed in our rebuttals to get this obvious point in the heads of those referees, I referred to in the above!

To give some idea how these numbers work, take characteristic dimensions associated with *E. coli*: $M \simeq 10^{-15}$ kg, $a, L \simeq 5 \times 10^{-6}$ m, $b \simeq 0.9a$ and it follows that $\tau_G \simeq 0.1$ seconds. This is the highlight prediction of Penrose, that triggered an intense experimental effort: create an object of the size of *E. coli*, put it in a spatial coherent superposition with itself, displaced by a couple of micrometres, and in a time span of the order of a second, one should find that this state spontaneously collapses. One would then like to study the collapse dynamics in great experimental detail, to get data on how this quantum gravity affair actually works. The above is not a theory, but instead no more than a dimensional analysis estimate of where the unitary business as usual may get defeated by the ambiguous time associated with gravity.

A standard objection is rooted in the standard semi-classical re-quantization logic, lifting classical field theory into quantum field theory. Unleashing this on (linearized) gravity, it follows immediately that any quantum effect should be strongly suppressed because of the smallness of Newton's constant G. Following the standard, semiclassical re-quantization recipe behind standard quantum field theory, only when 'radiative' corrections associated with emitting and absorbing gravitons start to play a role should quantum effects become discernible. As it was explained to me by Kip Thorne, after roughly a century, constantly waving your (human) arms, you will have emitted a single graviton. How can this ever influence quantum physics?

But the issue is that in this way, one brushes under the rug that one can no longer *use* the standard 'diagrammatics' of quantum field theory, given the difficulty unique to gravity, that the time axis no longer permits the unitary evolution, which is actually a prerequisite for the validity of the Feynman diagrams.

Obviously, the smallness of G does imply that the time dilation effects are very small. For instance, the Shapiro time delay associated with solar system dimensions (distance Earth–Mars, mass of the Sun) is in the millisecond range. Accordingly, the time difference in Tjerk's thought experiment involving *E. coli*'s dimensions is absurdly small, $t_a - t_b \simeq 10^{-35}$ seconds. How can this ever be of any significance?

But the issue is that now the *absolute* energy matters: this very short time is multiplied by the very large rest mass energy $Mc^2/\hbar$, combining in the dimensionless phase—this is responsible for the collapse time landing in the *E. coli* window.

One may plug in atomic dimensions—Angstrom length scales and the atomic mass: one finds out that τ_G time exceeds by many orders of magnitude the measured lower bound for the life time of the proton. According to this estimate, the unitarity of quantum physics on the microscopic scale is assured. This better be the case, since otherwise the quarks in your body (and so forth) would spontaneously collapse, turning you into a fireball. This is obviously not happening—empirically the 'unitary dance' of equilibrium microscopic matter lasts forever.

I perceive this as a crucial insight. Let us again collect all the pieces. It is an obvious fact that the macroscopic universe is governed by causal evolutions. This requires that the universe is in a non-equilibrium state, and gravity is uniquely responsible for this condition. This leads to an inhomogeneous distribution of stress-energy, which in turn implies the ambiguity of time: the ticking of the clock depends on the mass distributions—the Shapiro affair. The big deal is that when these time ambiguities are unnoticeable, the universe settles in the tranquil, 'timeless' unitary time evolutions of quantum physics. However, upon ascending to the macroscopic realm, the factor of rest *mass* becomes increasingly important, and thereby the 'pull of gravity'. Believing the dimensional analysis in the above, the time ambiguities wrecking unitarity become noticeable on the *E. coli* mesoscopic scale. Although we have no clue how this really works, an obvious net effect may be that unitarity comes to an end, specifically according to the Born rule book governing the collapse of the wavefunction.

One could then hope, when it would become possible to track in experiment precisely how this gravitational collapse evolves, one would obtain clues allowing us to reconstruct the 'quantum gravity theory' behind this collapse dynamics. This is a very different and way more optimistic outlook than the quantum gravity community folklore, that such clues can be obtained only by approaching the fundamental Planck scale. This requires impossible missions like inspecting the universe just after the big bang, or alternatively, charting the singularity in the middle of a black hole.

4.4 The stochastic evolution of thermal states and the gravitational wave function collapse

I have arrived at a point where I can close yet another circle, the one associated with this chapter. In Section 4.2 I reviewed the idea by D'Alessio *et al.* suggesting that the stochastic nature of the 'steam engine dynamics' in the finite temperature macroscopic world can be reconstructed, asserting that *repeated spontaneous wavefunction collapses* are at its origin. You are just done with my review in Section 4.3 of the notion of the *gravitational wavefunction collapse*. I am not aware of anybody else who has realized it—I do have another appealing no-brainer to offer. These two ideas complement each other in a striking fashion!

One may argue that the Penrose–Diosi ideas are in a way polluted by the 1930s tiny quantum systems syndrome. It is of course meaningful as a minimal set up aimed at observing the gravitational wavefunction collapse. But as the plain vanilla Schrödinger cat, it is not at all representative for how nature works in 'its natural state'. The key is in the observation that macroscopic matter *never* exhibits unitary time evolution, and Penrose argues that this is for the reason that gravity has forced the wavefunctions to collapse. But what is then governing the dynamics of macroscopic matter? We have known the answer since the nineteenth century: it is the stochastic affair called thermodynamics. And in Section 4.2 we learned that this can be impeccably reconstructed, asserting that repeated *spontaneous* collapses occur on an unspecified but short timescale, interluded by intervals of unitary evolution. However, D'Alessio *et al.* are silent regarding the origin of these spontaneous collapses.

Obviously, this implicitly assumes an 'objective' collapse mechanism: the guts of this affair is of course that these collapses happen automatically, without the intervention of observers. It is very obvious: we learn from the Penrose school of thought that *gravity* may be the culprit, while the dimensional analysis in Section 4.3 adds credibility to the notion that the characteristic collapse time may be of the desired 'short' magnitude.

In hindsight, it is also obvious to expect that after this collapse has taken place the large system re-enters an interval of unitary time evolution. There is no reason that I can imagine for a single collapse to completely destroy unitarity at later times. To the contrary, according to the Oosterkamp construction, collapses occur after a time *interval*: it is natural to assert that this implies that the collapses will occur repeatedly.

D'Alessio *et al.* also give an unambiguous answer to the question: what is the 'measurement basis' associated with the Born rule projections? In the Penrose–Diosi and the Oosterkamp 'cat set ups', this measurement basis is actually position space, for the reason that the system is prepared to collapse in this way by the experimentalist. But when the experimentalist is not preparing the system and observing the measurement outcomes, what to take for the basis that is, in an absolute way, preferred? As D'Alessio *et al.* demonstrate, in order to reconstruct the temperature-'driven' stochastic equations of thermodynamics this basis *has* to be the basis of energy eigenstates. As I already argued, although there are no a priori arguments available, it appears to be quite natural given that time is the culprit and energy is conjugate to time. For the non-physicist: the meaning of 'conjugate' is that eventually the Hamiltonian as energy operator governs the 'rotations' of the system in time in the Hilbert space, that we call unitary tim evolution.

Once again, the D'Alessio *et al.* and Penrose–Diosi appear to complement each other perfectly, actually even resolving some intrinsic ambiguities in the latter in a satisfactory way. I perceive it as a stunning insight: it yields a vivid and consistent insight into the *origin of heat itself*, as rooted in the fundamental gravity-quantum conflict. It insists that reality is therefore intrinsically stochastic, given the collapse being behind it, albeit in the time-tested sense of thermodynamics. This is good, since we know for sure that thermodynamics governs the causal evolutions of finite temperature macroscopic matter impeccably.

4.5 Observing the repeated spontaneous collapses

This is all qualitative. Can it be further quantified? Specifically, if correct it has to be that the spontaneous gravitational collapses happen all the time in literally *everything*. Thinking it over for a moment, one may arrive at the conclusion that these collapses are very hidden, shrouded from direct observation. It has to be like that since otherwise experimentalists would have observed them a long time ago. This should again be related to the difficult to measure intermediate *E. coli* scale, that will be in one or the other way still relevant also in this context. But how to estimate this scale more precisely, given that the conditions are quite different from the Diosi–Penrose 'cat'? Put more strongly, could it be that an experimental protocol can be figured out resting in one or the other way on 'gas clouds', that would unambiguously reveal the presence and detailed workings of the spontaneous collapses? Could it be that this may be eventually easier than the Schrödinger cat style experiments where one has to work extremely hard to fabricate the *E. coli*-style devices?

Frankly, I have not managed to nail this completely. When one tries to marry the D'Alessio *et al.* ideas with the gravitational wavefunction collapse, one meets new difficulties where I do not have resolutions to offer. In a first step, let us try to obtain order of magnitude estimates for the scales where the gravitational collapse may happen in typical thermodynamical set ups. The first difficulty I encountered is to find a *general* way of estimating this scale, regardless the details of a particular set up.

The interest should be in 'steam engine-like' circumstances. So much is clear: anything that is *static* will not exhibit the collapses; static means here of course equilibrium. On the other hand, any thermal system characterized by *currents* that flow will dissipate, and this heat production should eventually originate in the collapses. The only means that is available to address the gravitational wavefunction collapse is in the form of the Diosi–Penrose 'cat logic', in particular in the Oosterkamp incarnation. All I know to do is to pick thermodynamical circumstances, where I can discern how to employ this logic.

I figured out the following example. Consider a jet of gas formed by letting gas from a high-pressure reservoir escape through a nozzle into a low-pressure container. The particular nozzle of interest is the 'full cone spray nozzle' that yields a cone-shaped jet. This jet is an example of a dissipative current. Its expansion acts to cool the environment, but this is just due to the expansion. The entropy of the system will increase and in the closed-circuit architecture of a refrigerator, this expansion phase will contribute to the loss of work into heating the system.

The expanding jet may be itself stationary: in time the *density* distribution may be independent of time. According to the Diosi/Penrose/Oosterkamp logic, one somehow has to invoke different mass distributions that are in coherent superposition with each other, to trigger the gravitational collapse. How can this happen when the overall mass distribution looks the same as a function of time? A while ago we addressed this in another paper [21], specifically dealing with the potentiality of *flux qubits* as detectors of the gravitational collapse. The flux qubit is similar in the regard that it is characterized by a left and right chiral circular supercurrent forced in Schrödinger cat coherent superposition. If the mass distribution of both currents is identical, why should this then be subjected to the gravitational collapse?

We pointed out a simple principle that we found rather obvious: *it is a necessary condition for the coherence of the whole that all its parts are also coherent.* As an instructive metaphor we referred to a train. For the train to be coherent, of course all its cars should be coherent as well. Imagine a train covering completely a circular track speeding at a constant velocity v. Take a snap shot at time 'zero' and pick a particular car at position R_0. At a later time t, this car will have moved to $R = R_0 + vt$. This car represents the same gravitational body and since it has moved, the 'Shapiro time' will be different, signalling that it may collapse.

Read for 'car' the 'element of fluid', being the very fundament of hydrodynamics, and we may have an algorithm that may shed light on the collapse scale in thermodynamic settings. It seems that I can make this to work with the nozzle set up (it seems to fail for the simple pipe flow). Take as the first instance of time ($t = 0$), the volume of gas that is at that moment passing the nozzle. The nozzle opening has a radius R_0 and accordingly, we estimate this gas volume as a sphere with radius $V_0 = (4/3)\pi R_0^3$. The total mass of this sphere is $M = \rho V$ where ρ is the mass density. In addition, we assume that the jet is 'sonic', characterized by the (maximal) sound velocity v_s, of the order of 300 m/s for a gas at room temperature. We assume that this also governs the lateral expansion of the cone. This implies that after a time t, the gas sphere has expanded to a radius $R(t) = R_0 + v_s t$.

Having specified the way that mass distributions effectively change in time, we have now to cope with a crucial question. The Diosi/Penrose/Oosterkamp ploys depart from a Schrödinger cat set up, invoking quantum copies that are separated in space at locations that are however not changing in time. Such a condition has to be realized by the intervention of the experimentalist. Dealing with the spontaneous collapses there is of course no 'two-ness' that can be identified, and neither are there quantum copies that are at every instance of time separated in space. In whatever way, we are dealing with single massive bodies that do change their position in space during the temporal evolution.

It is obscure how to link this to the gravitational wavefunction collapse, just resting on the gravitational self-energy dimensional analysis, since this is just tailored to address specifically the static cat. The reader should have sensed that the Oosterkamp re-interpretation in terms of the Shapiro time delay is more general—the Diosi–Penrose cat is a special case. In the present context this greater flexibility becomes consequential. One can argue in the following way: in the course of the time evolution of a single gravitating body, the ambiguity in the definition of time quantified through the 'Oosterkamp clock' may grow because the body is changing its mass distribution in space. Instead of referring to static positions in the guise of the Diosi–Penrose cat, one may instead *determine the change in the Shapiro time delays as a function of the temporal evolution of the spatial mass distributions.* The gravitational collapse time can then be estimated as before, as the scale where the initial and later time wavefunctions acquire a gravitational phase shift (Eq. 4.9) becoming of order 2π.

The reader may now guess why I came up with the cone nozzle ploy. In combination with the 'train-car principle' we can address the collapse time directly in terms of the small- and large-sphere cat state formulas of Oosterkamp. The initial time is associated with the gas volume on the verge of leaving the nozzle (sphere with radius R_0) and at the late (collapse) time this sphere has expanded into the larger $R(t)$, while the total mass has not changed. Hence, we can now use directly Eq.(4.10) with $a = L = R_0$ and $b = R(t) = R_0 + v_s t$, finding for the collapse time,

$$\tau_G^{jet} = \frac{4\pi\hbar R_0^2}{8GM^2 v_s \tau_G^{jet}} \tag{4.11}$$

But since we asserted that the expansion of the radius of the sphere is determined by time $\sim v_s \tau_G^{jet}$ we obtain an estimate for τ_G^{jet} that only depends on the system parameters ρ, R_0 and v_s,

$$\tau_G^{jet} = \frac{1}{\rho R_0^2} \sqrt{\frac{9}{32\pi} \frac{\hbar}{G v_s}} \tag{4.12}$$

Let us estimate the order of magnitude of this collapse time. For a typical gas under ambient conditions, the order of magnitude of the mass density $\rho \sim 1 \text{kg/m}^3$, while for the sound velocity $v_s \sim 10^2 \text{m/s}$. Assume a nozzle radius $R_0 \sim 10^{-3} m$ and we find $\tau_G = 0.1$ microseconds!

In a way this time is remarkably mundane. Although still very short for humans it is snail pace for electronic lab equipment. In the above jet nozzle set up, one could expect the signature(s) of the wavefunction collapse at a distance $L_G \simeq 10$ micrometres from the nozzle. This may be tricky since this is still small on the scale of the dimension of the nozzle itself, but it is surely promising that things can be done by a clever experimentalist.

The trouble is, however, what do we expect to happen at τ_G^{jet}, L_G? There should be obvious signatures, dealing with a strictly non-interacting quantum gas. Regardless whether the particles are fermions or bosons or whether the temperature is high or low, the states are formed from the single particle quantum mechanical eigenstates. These are the plane waves of the particle in the box exercises, and since these are waves, they will interfere. One may now conceive an arrangement placing particle mirrors surrounding the nozzle. The particle waves will partly reflect and interfere with each other, giving rise to interference fringes as long as the system submits to the unitary evolution. At the instant of the wavefunction collapse, they turn into classical particles and the interference phenomena will vanish.

However, non-interacting particles do not exist. The interactions in a gas are captured by the time it takes to collide with another particle, the mean free time and associated mean free path. The mean free time τ_{coll} of a gas under ambient conditions is of the order of a nanosecond (10^{-9} seconds). We find that $\tau_G^{jet} \simeq 100\tau_{coll}$. In the kinetic gas, this mean free time heralds the onset of the *quantum chaos* in the sense we discussed: at times longer than τ_{coll} the wavefunction starts to spread into the exponentially large Hilbert space. This is actually a beneficial situation for the mapping on the classical double stochastic dynamics in the D'Alessio *et al.* scenario, set by the repeated collapses since this actually is critically dependent on being in a quantum chaotic regime, when the collapse occurs.

However, how much observable difference is there to be expected, pending the wavefunction being collapsed or not when the system is in this quantum chaotic regime? Surely, signatures like the interference fringes of the independent particles will be completely washed away by the collisions, since single particle momentum (the quantum number associated with the plane waves) is no longer conserved. This is the first problem I have encountered with this ploy: frankly, I have no clue what to look for in such a 'steam engine' style experiment, revealing the presence of the spontaneous wavefunction collapse. This despite the fact that, at least in this nozzle experiment, the fundamental gravitational collapse time may be within easy reach.

But I have more headaches. I picked the nozzle set up for the reason that it is rather closely related to the Oosterkamp ploy involving spheres with different volumes and densities. This is however a rather special affair dealing more generally with dissipative thermodynamical phenomena. Perhaps the simplest circumstance is the low Reynolds number flow of a viscous fluid through a pipe. Again resting on the train-car metaphor one can now address the elements of fluid that are relevant in this context that represent a *static* mass in the direction transversal to the flow. In the first instance these do not change, dealing with the current flowing in the direction of the pipe. But in this 'longitudinal' direction we can again employ the 'modified' Oosterkamp Shapiro time estimate. We actually considered such a situation dealing with the gravitational collapse of counter-propagating supercurrents in a flux qubit. Using the gravitational self-energy estimate the collapse time is given by Eq. (4.11) in Ref. [21]

$$\tau_G^{pipe} \simeq \sqrt{\frac{\hbar}{G(\rho A)^2 v}} \qquad (4.13)$$

where A is the cross-sectional area of the pipe and v the velocity of the current. This scales $\sim 1/A$, implying that pending the pipe diameter τ_G^{pipe}, it can vary all over the place. Assuming a radius of a micrometre, as for the flux qubits, this may well be in range of macroscopic times. However, assuming the pipe diameter to be macroscopic, say 10 cm, one obtains times of the order of nanoseconds, already deeply into the inconvenient microscopic range. But the real difficulty I have is in the degree of arbitrariness that appears to be intrinsic to this style of dimensional analysis, where the outcomes are completely different, pending the details of the macroscopic set up as the above examples demonstrate. The reader is cordially invited to dream up something more satisfactory.

The greatest difficulty is still to come. I already stressed that in order for the collapses to occur repeatedly in the energy basis, as required for the mapping on the double stochastic dynamics, a parametric time dependence of the Hamiltonian is required. To reconstruct the double stochastic dynamics, the state after a collapse time step should be stochastically distributed between the different energy eigenstates associated with the Hamiltonian realized at that particular time. However, when the Hamiltonian is not changing in the interval preceding this particular collapse, the system will continue to be in a state *identical* to the one realized at the preceding collapse. The energy basis continues to be the same and at the preceding collapse, the system has already been projected on this basis.

This seems to introduce another sense of arbitrariness. Thermodynamic evolutions typically invoke forcing: consider for instance the basic thermodynamic set up of gas in a cylinder that is compressed by applying an external force on a piston. This will indeed be captured by a parametric time dependence in the Hamiltonian that may be of the right kind for the D'Alessio *et al.* repeated collapse mechanism. In this regard one may even speculate that this external time dependence may actually take the role of a bottleneck. It is imaginable that the characteristic time, associated with the external drive, changing the Hamiltonian sufficiently for the stochastic 'shuffle' to happen, may be longer than τ_G. In such a case, many collapses may happen without reshuffling, until this external time takes over.

The trouble I have is, however, with *quench*-type experiments. One considers here a Hamiltonian that is time independent for 'negative' times, to suddenly change at time 'zero' in a different Hamiltonian that is again time independent at positive times. Such quench experiments are quite popular among practitioners of quantum non-equilibrium because of the relative ease to compute matters. Such events are maximally anti-adiabatic because of the sudden nature of the change and are therefore maximizing entropy production/dissipation. The nozzle set up may be an example. At negative times, one has a sealed vessel at a high pressure in a low-pressure environment. At time zero, one just opens a valve leading to the noz-zle and at positive times the high-pressure vessel is slowly losing the gas without any external force acting on it.

But this means that the positive time Hamiltonian of interest should be time independent! How can this be related to the D'Alessio *et al.* repeated collapses, while we know that the macroscopic flows associated with the nozzle are impeccably captured by thermodynamics? Frankly, I have no clue.

5

The quantum gravity mainstream versus my problem of time

I am nearing the end of what I wanted to tell. Obviously, this whole story has been bouncing back and forth between the two triumphant theories of modern physics, general relativity and quantum (field) theory. I highlighted the tension between these theories that invariably revolved in my story around the dimension of time. These can be summarized in the three 'no-brainers', that I perceive as very obvious, but by and large not recognized up to now by mankind: (1) Gravity as the cradle of causality (Chapter 2), (2) The trouble of quantum physics dealing with causality (Chapter 3), and (3) Thermodynamics as a consequence of repeated spontaneous collapses which are in turn rooted in the gravitational wave function collapse (Chapter 4).

This is surely somehow addressing the relationship between gravity (GR) and the 'quantum'. But how does it relate to the mainstream thinking about this relationship? There appears to be a full consensus among physicists that the problem of quantum gravity is the number one problem in fundamental physics. However, this is for yet rather different reasons than my list in the previous paragraph. These are also very good reasons.

The most violent collision between the two theories is here in the limelight. Upon ascending to higher and higher energy it should be that the 'quantum' becomes increasingly manifest in dealing with gravity. Following the usual semi-classical route of re-quantizing the classical theory as explained in Section 3.1 (by weighing the world histories with the classical action), one finds for GR that the quantum corrections become increasingly important upon descending to shorter times and distances, corresponding with higher energies. These actually get completely out of hand: GR is 'non-renormalizable' which means that the classical low-energy theory has completely forgotten how the high-energy theory works. All one has is dimensional analysis. Classical GR has actually no knowledge of absolute scales; it is left unchanged by global changes of units. However, an absolute scale falls out when Newton's constant G and the velocity of light c are combined with $\hbar$. One can isolate a characteristic 'Planck length', $l_P = \sqrt{\hbar G/c^3}$, a 'Planck time' $t_P = \sqrt{\hbar G/c^5}$, a 'Planck mass' $m_P = \sqrt{\hbar c/G}$, and so forth.

It has to be that at this scale that the quantum effects on gravity and the gravity effects on the quantum should be on centre stage. But the Planck scale energy is extremely large, $\simeq 10^{16}$ TeV (Tera electron Volt, a million mega electron

On Time. Jan Zaanen, Oxford University Press. © Oxford University Press (2024).
DOI: 10.1093/9780198920793.003.0005

Volts). The most powerful accelerator is presently the Large Hadron Collider of CERN in Geneva, reaching a maximum energy of 13.6 TeV. Hence, this Planck-scale physics is completely out of the range of observable physics. Yet, in the most extreme gravitational circumstances embodied by the cosmological (Big Bang) and black hole singularities, this Planck-scale quantum gravity should take over.

The number one task of mainstream quantum gravity as it has evolved over the last half a century, has been to address the nature of this Planck-scale quantum gravity. Lacking any experimental information, this has been entirely propelled by mathematical means. This includes ventures like the 'loop quantum gravity' and the 'causal dynamical triangulation', characteristically focussed on getting a handle on a quantized, somehow discrete geometry. But the largest community that may claim the greatest progress is formed by the string theorists.

String theory started in the high-energy quantum field theory tradition and this heritage is felt until the present day. It started in the 1970s as the quantum theory of relativistic 'line-like' objects in space—the strings. Then remarkable mathematical miracles start to happen—I took the graduate course and in lecture four or so you find yourself busy with solving Einstein equations! There is just a deep and to my perception quite mysterious mathematical connection between quantum theory and general relativity that becomes manifest in this context.

All along, this was perceived as somehow shedding light on the Planck-scale questions, and a remarkable development followed over a 40-year or so period, propelled by the mathematics, delivering interesting results one after the other. This eventually culminated in the AdS/CFT correspondence, the 'holographic duality' that I have already alluded to a number of times.

5.1 The glory of string theory: the holographic duality

This is just a case in point with regard to mathematical relations between general relativity and quantum theory. As a matter of fact I have myself been an enthusiastic user of the 'correspondence' during the last 15 years. This is a quite pragmatic affair. My interest is in the observable properties of extremely strongly interacting, densely many-body entangled matter that lives 'behind the sign problem brick wall' (explained in Section 3.1). In recent years the focus has been particularly on the quantum thermalization associated with non-equilibrium circumstances. The correspondence is just acting as a calculator, translating it to a general relativity problem in a space with an extra dimension, that can be solved in principle, although it may stretch the limits of GR mathematical technology. In a wonderful way, the correspondence captures quantum physics, when it is in the domain of exponential complexity, that cannot be computed in any other way, pending the arrival of the quantum computer.

This has grown into a vast subject by itself and I will keep quiet about it in the present context—I have been responsible for plenty of other introductory texts [8; 7]. But on the most basic level, the correspondence reflects the underlying paradigm of the string theorists. String theory is a continuation of the quantum field theory tradition, and it just takes the standard formulation of quantum theory with its unitary sector and Born rule collapse for granted. The questions I have been raising in the form of the no-brainers are just ignored.

Intriguingly, the correspondence works in such a way that it intrinsically avoids the troubles that at least I expect, dealing with the mixture of gravity and quantum physics. The reason is that the gravity and quantum sides are in a literal sense realized in different, disconnected worlds. Potentially there is a lot more to it, but the part that is established, delivering for my purposes, consists of a so-called 'bulk': a 'universe' with the extra dimension that is governed by strictly *classical* gravity/general relativity. This then relates to the 'boundary', with one less dimension, that describes a non-gravitating flat geometry space-time, on which the extremely strongly coupled quantum theory is realized. This is just part of the quantum field theory agenda, with the difference that with the correspondence we can compute matters that are beyond the reach of the conventional field-theoretical machinery.

These two realities are in 'dual relation'. Dualities are kind of magical mathematical devices. These are of a variety of types, but they share a conceptual fundament, that may be formulated as: 'opposites belong together, forming a harmonious wholeness that appear, pending the questions one asks'. You may be familiar with the Yin–Yang principle of Taoist metaphysics. I got this explained by Chinese friends—with tongue in cheek—as follows. Yin is cold and Yang is hot. Yin is cat and Yang is dog. Therefore, in order to restore the harmonious balance, living in the north of China close to the Korean border you should eat *dog* but when you live in the south near Guangzhou you have instead to eat *cat*.

A first example of duality in physics is the *particle–wave* duality of single particle quantum mechanics: pending how one measures the system, the particle may behave as a cannon ball or as a diffracting wave. The mathematical machine behind, is actually the Fourier transformation that may be familiar to the mathematically literate reader. AdS/CFT is in a way the ultimate duality—I haven't encountered anything that excels to the same degree in relating 'opposite' physical universe in such an astonishing richness. I conceived yet another Yin–Yang joke that expresses this sentiment. 'A long time ago God invited a Taoist monk to heaven. The monk asked God regarding the secret of existence and God responded by explaining the correspondence. But God speaks with the tongue of Edward Witten—the monk found it all crystal clear, but when he was back among the mortals all he managed to reconstruct is the cat and dog duality affair'.

The string theoretical prophet Edward Witten is among others famous for his extremely transparent presentations that may be hard to reconstruct after the fact...

The take-home message is that the time troubles associated with quantum and gravity just disappear because gravity is just the consistent, causal, and so forth classical incarnation living on one side of the duality. On the other hand, the quantum theory in standard formulation is devoid of the difficulties, since its place is non-gravitating. To every 'event' in the quantum world there is a corresponding image in the GR world and vice versa. This is governed by a set of tight rules called the dictionary. For instance, a finite temperature state in the boundary requires a black hole with a finite 'thermal' horizon in the bulk. In fact, this directly employs the Euclidean time circle of the black hole, highlighted in Section 2.5. As it turns out, the space-time of the boundary quantum theory lives at the radial infinity of the bulk. Although the bulk is strictly classical, the boundary quantum theory picks this up as the finite temperature equilibrium state, generic for a quantum field theory as explained in Section 3.4.

In this very clever way, the correspondence avoids my troubles with time in the gravity–quantum mixture. The correspondence also excels in addressing non-equilibrium physics, but this follows the established rule book. One first decides how to incorporate the 'by hand' parametric time dependence in the boundary. This then translates, using the dictionary, into a time-dependent protocol for the classical (Einstein) equations of motion in the bulk. Translating the solutions to the boundary then yields often surprising, but very reasonable and correct, non-equilibrium quantum phenomena in the boundary. For instance, recently I had a good time with Chinese collaborators, computing the way that the quantized vortices of superfluids and superconductors dissipate their motional energy [22], an affair where one really needs to address matters in terms of the underlying quantum theory.

This is not all, since the question arises how the correspondence deals with the collapse of the wavefunction. I got in the course of time, increasingly intrigued by an aspect that seems to be taken for granted by the string theorists, that I perceive however as not so self-evident. *Insofar as the correspondence has delivered tangible results, these pertain entirely to the behaviour of observables, i.e. the VEVs that remain after the collapse of the wavefunction.* Somehow, the information on the unitary sector is completely hidden; there is nothing in the bulk telling us anything regarding this part of the quantum dynamics. A crisp way of expressing this fundamental deficit, is in the realization that the quantum states processed by the correspondence are characterized by exponential complexity, delocalized in the extremely large many-body Hilbert space [7]. On the gravity side, one encounters the low information density gravity solutions that may fit on the back of an envelope. But these are fine to describe the VEVs; the observable responses of the quantum theory may be very simple, especially so dealing with an exponential complexity of the unitary evolution [7].

It is argued by the string theorists that one can discern an underlying 'M-theory'—this is quite remarkable, as there is evidence for the existence of such a theory, but it is totally in the dark what the theory is. It has also been argued that the correspondence is implied by M-theory, although the part that is explicitly available is tied to a special limit—the classical GR bulk. It makes me wonder, could it be that M-theory knows about gravitational wavefunction collapses, thereby insisting that the macroscopic matter described in the boundary, is entirely about expectation values?

5.2 A critique on Hawking I: the Unruh effect

If there is a set of equations that I completely trust, it is the AdS/CFT correspondence. The question is whether nature wants to play its games, but by itself it is of an astonishing correctness. It is, however, in the string theory culture, intertwined with a much older school of thought, when it comes to quantum gravity aspects. This is the paradigm that emerged in 1974 by Hawking's construction of his black hole radiation. I already discussed this shortly in Section 3.4. Once again, the claim is that a black hole turns into a material object, characterized by a (Bekenstein–Hawking) entropy, and a black body radiation associated with the Hawking temperature, caused by the black hole geometry interfering with the zero point quantum motions intrinsic to the quantum field theoretical vacuum. You already met the simplest way to derive this, in the form of the Euclidean time circle at radial infinity, associated with the Schwarzschild metric of the eternal black hole (Sections 2.5 and 3.4).

This actually played a key role in the development leading to Maldacena's discovery of the correspondence in 1997. I already discussed in the above, how the thermal properties of the boundary theory are encoded in the bulk geometry—it is actually coincident with the Hawking results. AdS/CFT is also called a 'holographic duality': the boundary has one less dimension than the bulk. This is like a two-dimensional holographic screen, showing complicated interference patterns (like the quantum theory). By shining a laser through it, a tangible three-dimensional image appears (like the bulk black hole). All along it was clear that the Bekenstein–Hawking black hole entropy scales with the *area* of the event horizon—very different from normal material objects, having an entropy scaling like the *volume*. In the early 1990s, 't Hooft and Susskind formulated this as the *holographic principle*: (quantum) gravitational theories, as compared to the 'material' field theoretical realities, have fewer degrees of freedom and these can be counted by removing a spatial dimension.

Together with the specific solutions for the temperature, entropy, and so forth, this holographic principle played a key role in getting AdS/CFT on the right track. But I discern a deeply rooted confusion in this community, that is to the best of my understanding, a product of this historical development.

The Hawking radiation affair departs from a black hole space-time, living *together* with a free-field quantum vacuum. However, in AdS/CFT the quantum part is strongly interacting but non-gravitational, while the gravity part is strictly classical, and they are just talking to each other by the mathematical duality mapping. It appears that in the string theory tradition, these very different situations are mixed up, treated as one and the same thing, when it comes to these quantum gravity aspects. To get an impression, one may wish to Google a presently fashionable subject in this community labelled by 'Page curve' and 'islands'. The 'Page curve' refers to fanciful follow-up work on Hawking radiation, while the 'islands' refer to a particular gravitational constellation realized in the holographic bulk, believed to encode for this.

The specific issue I have with these matters is that I discern a variety of quite hairy issues with the 'Hawking way', which are somewhat magically avoided in the correspondence. These are shrouded in the main stream way of thinking about these matters by the rather intuitive appeal exerted by the Hawking idea of black holes as thermal objects. In more than one way, one cannot trust the underlying mathematics. In the next chapter, I will focus in on the uncontrolled assumptions involved in the information paradox. But let me first address the essence of the Hawking mechanism in the form of the Unruh effect. This Unruh effect is just signalling yet another tension between quantum theory and general relativity. The way this is resolved in the standard 'Hawking folklore' may be correct, but it surely does not follow unambiguously from quantum theory and/or gravity. It actually involves implicitly an extra postulate that is not obvious *a priori*, while it is brushed under the rug in the community folklore.

The Unruh effect is yet another simple way to understand the origin of Hawking radiation. Given the equivalence principle—inertial and gravitational mass are the same thing—it is just impossible to discern the difference between experiencing gravitational attraction or being accelerated by the same force. One can easily deduce the strength of the 'surface gravity', the gravitational field near the event horizon of the black hole, defined as the acceleration exerted at infinity required to keep an object at the horizon.

One can now construct a coordinate system as of relevance to a probe observer, who is however equipped with a rocket pack that accelerates the observer so that it hovers at a fixed distance from the horizon: the (Cartesian) Rindler coordinates. Such an observer will experience a ('Rindler') event horizon, which is coincident with the black hole horizon as implied by the equivalence principle. Following the same logic, such an observer that is falling freely to the black hole since he/she did not ignite their rockets, will experience locally just a flat Minkowski geometry—the event horizon is a so-called coordinate singularity and only the observer hovering near the horizon will experience it in the Rindler way.

Given this simple insight we can now just forget the black hole: this horizon is just a property of an accelerating frame, and one may just accelerate using only a rocket pack. But given the horizon, it is impossible for the observer to harvest any form of information from behind the horizon. Dealing with the zero-point motion style quantum fields populating this space-time, it appears that one should assert that the quantum fields residing behind the horizon have to be traced out from the total density matrix as of relevance to the observer. This is the crucial *extra postulate* in the original Hawking argument. The resulting reduced density matrix is the one governing the quantum reality experienced by the rocket pack observer. This turns out to be a thermal density matrix characterized by the temperature $k_B T_U = \frac{\hbar a}{2\pi c}$ where a and c are the acceleration and the velocity of light, respectively. This is then in turn precisely coincident with the temperature found employing the Euclidean time circle of Sections 2.5 and 3.4.

The take-home message is that, even when you push the gas pedal of your car, the temperature in your environment is automatically rising because of this logic. But even for the most powerful car, this temperature increase is so extremely tiny that it cannot be measured. Black holes are just special because of the numbers—one needs fantastic rockets that cannot be built to hover near their horizon.

This is the standard intuition in the community behind the Hawking radiation. However, there is a critical mathematical problem that is shuffled under the rug. The foundation of general relativity rests on 'general covariance' as Einstein called it, while mathematicians call it 'diffeomorphism invariance'. This is just insisting that the GR solutions cannot possibly depend on the choice of coordinate system. After all, coordinate systems are no more than a convenience for the calculations made up by humans, and not a property of space-time. The transformation from Minkowski (flat) coordinates to Rindler coordinates is no more than a *coordinate transformation* and the outcome of the theory cannot depend on the choice. This is a truism in classical GR.

Let us now turn to the Unruh effect. The precise mathematical statement is that upon a mere coordinate transformation from a flat (Minkowski) to an accelerating (Rindler) frame, the *physics is altered!* The zero temperature state in the flat frame becomes a finite temperature thermal state in the Rindler frame. I find this an alarming affair. How can we trust the equations when such a breach of seemingly defining principle (general covariance) is at work?

Notice that this is an issue which is unrelated to the 'Penrosian conflict' associated with the lack of global time in a back-reacted GR geometry. The Rindler frame is a way to express the Schwarzschild black hole geometry and I stressed that this is a stationary affair which is therefore characterized by an unambiguous global time. In fact, on the unitary evolution level there is a sound unitary transformation, keeping this part of the problem invariant under the coordinate transformation. But this fouls up upon mobilizing the notion of an observer, that has only access to information on his/her side of the event horizon. In the spirit of the Born rules, it is then asserted that the causally disconnected part of the quantum fields should be traced out: the natural incarnation of projective 'measurement', dealing with a density matrix associated with the state of the vacuum.

Although quite reasonable, it is an additional assumption, a separate postulate. How can one be sure that this *has* to be correct? It is not quite the same act as captured by observers submitting to the time-tested Born rules. Once again, the ramifications associated with sacrificing general covariance are quite confusing. Its most famous ramification is called 'the firewall' in the recent literature. This revolves around *the* consequence of this affair. On the one hand, consider an observer that is falling freely towards the black hole. Such an observer does not feel any force/acceleration: upon passing the horizon its frame continues to be flat and there is no such thing as an Unruh/Hawking temperature. However, we can equip the observer with a rocket pack, that allows him/her to hover arbitrarily close to the horizon. Upon approaching the horizon, the Unruh/Hawking temperature is increasing, diverging to infinity upon hitting the Schwarzschild radius.

How to reconcile this with the notion that there should be one reality? This kept very creative minds busy during the history of the subject. One way out is the idea of 'black hole complementarity' by Susskind, Thoracius, and 't Hooft, asserting that for causal reasons these two observers are incapable of exchanging information regarding the differences in what they observe. But this creates an inconsistency with the evolution of the entanglement of particles on both sides of the horizon. The 'firewall' refers to a 2012 proposal where this paradox is resolved by actually explicitly sacrificing the equivalence principle, asserting that there is actually a physical extremely hot 'membrane' (firewall) localized at the horizon. This idea attracted a lot of flak and the debate regarding this paradox is still continued.

My personal opinion is that given these deep difficulties, in combination with the ad hoc ('tracing out') assumption from where it departs, one better be quite agnostic. Does Hawking radiation exist? I am not entirely convinced, given the untested fundaments of the mathematical arguments. Keep in mind that there is up to the present day zero experimental support for it. The only mathematical construction that I perceive as watertight is, once again, the imaginary time circle associated with the Euclidean Schwarzschild geometry. One sees here the 'firewall' at work—the time circle shrinks to zero upon approaching the Schwarzschild radius, indicating that quantum fields would find this an infinitely high temperature place.

But this is supposed to be equilibrium physics, while the shrinking time circle of this Schwarzschild universe should then be characterized by steep temperature gradients. Thermodynamics seems to insist that this should then source heat currents that should diminish the temperature differences. By principle, these should be dissipative but nothing of the kind can be identified in this equilibrium Schwarzschild universe! I perceive it as a weird affair and I have not figured how it can be interpreted to make sense. This mirrors the firewall difficulties in the Lorentzian signature language, that I just discussed.

Yet again, *in AdS/CFT these difficulties disappear like snow before the Sun*. The gravitational bulk is strictly classical and in the bulk there are no quantum fields, for instance translating Euclidean time circles in temperature. This only matters at radial infinity where the radial asymptote of the time circle hits the QFT universe, devoid of gravity living in the boundary. This just picks up the Hawking temperature at radial infinity and this is devoid of any paradox.

More generally, I sometimes wonder whether this whole Hawking business, with its free quantum fields living in classical GR geometry, has just been a misleading coincidence. As I stressed, the portfolio of Hawking phenomena has a precise image in holography. But in holography there is the twist that 'the quantum' and 'the gravity' are strictly separated, only communicating by the duality mapping of the dictionary. In this way, basic wisdoms such as 'a black hole has an entropy' are fine, and only having the meaning of the entropy of the dual field theory in the boundary. The trouble may be that the classical limit of *quantum* gravity may be *singular*. This in the sense that any finite degree of 'hbar' in the gravitational sector may give rise to a 'quantum geometry' that is qualitatively different from the classical GR geometry, even to the degree that when this hbar becomes extremely small, the quantum geometry is still subtly different, impacting however on matters like the firewall. But as I announced in the very beginning, as everybody else, I have not the faintest clue how this quantum geometry works. I lack any form of ambition in this regard, other than that I find it a good idea to take nothing for granted.

5.3 A critique on Hawking II: the black hole information paradox

Arguably, the subject associated with the Hawking radiation agenda that has attracted most effort and debate over time, is the black hole information paradox. I have mostly been discussing eternal black holes, but let us focus in instead on the black holes that are by now well established by the recent gravitational wave and event horizon telescope observations to exist in our cosmos. These spring into existence typically by the collapse of a very heavy star at the end of its life. However, believing the Hawking mechanism, it will lose energy steadily due to the Hawking radiation. Although it will take an incredibly long time for an astrophysical black hole, eventually the black hole will evaporate again.

One can now throw an object containing information into the black hole. Imagine that the Encyclopedia Britannica' is swallowed up by the black hole, disappearing behind the horizon. Can this information be retrieved from the remnants of the black hole after it has completely evaporated? Our classical intuition would suggest that the black hole is more destructive for information than a paper shredder, and there does not seem to be much of an issue.

However, let us now consider this affair as a quantum process. Quantum should now be read as 'real quantum' meaning that the system as a whole, including the quantum version of the Encyclopedia, the black hole, and whatever else exists in this space-time, submits in the first instance to the *unitary* time evolution. Throw a unitary quantum mechanical wave into the black hole—the infalling state is a pure state. Given the 'trace out the stuff behind the horizon' prescription of Hawking, the outgoing state should be naively a mixed, thermal state. However, this violates the unitary time evolution, which insists that a pure state can only evolve into a pure state.

The crucial ingredient is now that the black hole will eventually evaporate. After this has happened, there is no longer a horizon. Isn't it the case that upon considering the whole time span between the appearance and disappearance of the black hole, it is eventually governed by unitary evolution when the system as a whole is considered? This should then imply that the quantum information absorbed by the black hole should re-emerge after the evaporation of the black hole in one or the other form.

This 'information paradox' triggered quite some upheaval in the community, among others in the form of bets between key players like Hawking and Preskill. In the course of time, this evolved into a community consensus that unitarity wins. The quantum information does not get lost and it is released in some form when eventually the black hole completely evaporates by the Hawking radiation. After all, the thermal nature of the exterior region can be traced back to the entanglement of the quantum degrees of freedom in the interior and exterior regions of the black hole geometry. When the horizon has disappeared, this entanglement protected by the unitary evolution becomes manifest again everywhere.

Frankly, I do find this strongly misguided. It is a typical example of a whole community suffering from tunnel vision. It is of course a tunnel shaped from fanciful mathematics, so it is very smart company. But maths, detached from reality—there is no empirical anchoring—can be quite delusional.

A genuine no-brainer is completely overlooked. Nobody has ever observed a *macroscopic* object that is exhibiting any sign of unitary time evolution. Black holes, as well as the Encyclopedia Britannica are *macroscopic objects*. Especially, black holes are typically very macroscopic, such as the supermassive black holes in the centre of galaxies that are of the order 10^{10} heavier than the Sun. I discussed in Chapter 4 at length why this unitarity has to come to an end, presenting the arguments pointing at the conflict with gravity to be the culprit. It is a very basic affair, and regardless whether one takes the Hawking mechanism of Section 5.2 for granted, it will interfere. When a black hole swallows up the encyclopedia, the mass distribution in this universe will change. The encyclopedia falls through the horizon, and since the mass of the black hole itself increases, its Schwarzschild radius will also change. There is no longer a global time-like Killing vector conflicting with the conditions for unitary time evolution as explained in Section 4.3.

Although we do not know why, we just know empirically that in the macroscopic domain, unitary time evolution comes to an end, including the notion of stationarity of information. To claim that after the black hole evaporation, somehow all the 'pure quantum info' will resurface again is as absurd as the idea that when my body gets cremated after I have passed away, all the thoughts that I accumulated in this text will escape in a little balloon or whatever, from the oven. In the big world information is all the time irretrievably destroyed—it is a principle called the second law.

Why should black holes be exempted? In fact, it turns out that in the community busy with the information paradox, firewalls, page curves, and other ramifications of 'black hole unitarity' (nowadays mostly string theorists), there is a near to complete lack of interest in the difficulties with time that have been the main theme of my musings. To the best of my understanding this is for historical reasons. This present day string theory tradition was born in the 1980s as a continuation of the high-energy particle physics tradition, resting on quantum field theory. In this context, the problem of time is much more implicit: one can address anything that matters in the standard formulation of quantum physics, with its split in unitary evolution followed by Born rule collapses. As I highlighted in Section 5.1, string theory has been in its own way very successful in relating certain aspects of the quantum reality to gravity. But insofar as this rests on trustworthy mathematical machinery, it meticulously avoids my troubles with time.

The awareness of these difficulties is lacking in this community, because it is too easy to look away from it. This is fine up to the point that equations are no longer used as the primary compass, while instead the arguments become of an intuitive, philosophical kind. Overlooking parts of reality, but also groupthink or even plain vanilla dogmatism then become considerable hazards, as the history of mankind demonstrates. Once again, I have a very high regard for the intellectual standards of the string theory community, but I am actually not confident that *any* string theorist will read these particular words. I do fear that even these most intelligent examples of the human race may already have thrown away this text long before this point, being subjected to an ideologically inspired rage.

6

Epilogue

I am done with the physics story I wanted to tell. But I have yet one more story in the offering, having as its only relation to the remainder of this text that it has to do with information while it revolves around yet another no-brainer. This one is, however, way less serious. It is not anchored in the equations of physics and it has quite a wishful thinking appeal. I am just floating an idea, hoping that you find it entertaining.

6.1 The eerie appeal of story telling

You know about the book 'Sapiens' by the philosopher Yuval Noah Harari, perhaps [23]? It is strongly recommended, as it is the most interesting line of thought regarding the nature of mankind that I have encountered in a long time. It revolves around a physics quality no-brainer. Like the equivalence principle, when realized it is completely obvious but somebody has to spell it out first. It is the capacity of the human race for *story telling*.

What is a story? What you read right now is a story. It is surely a collection of data, information that you can count in terms of the number of bits needed to encode it. But Harari means really a *good* story. It is something that only exists in our imagination while in order to be a good story it has to be original. This whole text has been intended to be a good story in this sense, including this last part. I will stitch together several established patterns 'inside the domain' to then arrive at a new synthesis that may well enrage you because you see it for the first time. Crucially, artificial 'intelligence' as represented by e.g. ChatGPT would for sure *never* 'get the idea' all by itself.

Humans are all the time engaged in story telling and it seems a completely self-evident activity. But this is deceptive.

How is it with other mammals? I have been worshipping intense friendships, and even love affairs, with a number of dogs and especially cats. With regard to the cats, a long while ago I discovered the 'eastern' siamese-like cats. They are much more social animals than the European variety. I find them actually superior to humans with regard to their emotional intelligence. I am worshipping presently an intense affair with Lola, an oriental short hair lady. I have given her full control over the maintenance of our love affair and she is remarkable, operating

On Time. Jan Zaanen, Oxford University Press. © Oxford University Press (2024).
DOI: 10.1093/9780198920793.003.0006

with astonishing effectiveness, avoiding routine, and keeping it maximally intense. However, can Lola tell a story? No way, this is where she fails and this is also true considering my dog friends.

More than anything else, this story-telling capacity makes us humans. Harari then argues that apparently this story-telling capacity showed up for the first time some 70,000 years ago; the 'cognitive revolution'. Stories can be exchanged and shared by groups of humans. In the process, matters that only exist in imagination—the 'substance' of stories—may become bigger than physical reality in the community of people. Grand narratives that developed in this way historically are for instance Gods, but also organizations (governments with presidents), human rights, and most strikingly money: in short about everything that defines human society. All these things do not exist physically, these are fruits of imagination that became bigger than life by a developing consensus, a hyper-conformism. More than anything else, these are the key to the success of mankind, since stories like money made it possible to organize mankind tightly and efficiently. There is a lot more to it and I recommend Harari's book to get exposed to the full glory of his case.

6.2 Why artificial intelligence is stupid

I discern here a connection to another recent affair—it seems that nobody else has noticed, although I perceive it as quite obvious. Presently the artificial intelligence hype is raging. Undoubtedly there is a potential to make a lot of money and the IT oligarchs (Google and so forth) have in this regard a rational investment strategy. It is a proven 'engineering tool' when it comes to routine pattern recognition. This may well imply that a lot of quasi-intellectual activity, as exhibited for instance by lawyers, medical doctors, and even journalists (relying on ChatGPT) can be taken over by computers. The clue is that these do not involve original thoughts, but instead rely on pattern recognition based on established precedents.

This may then get confused with real intellectual activity, the doom stories that AI machinery will become more intelligent than humans, taking over the planet in a Terminator-like doomsday scenario. There is however a glass ceiling to AI—I became acutely aware of it during a colloquium talk by a computer science professor from Amsterdam, who is considered to be the number one AI academic researcher in the Netherlands. He worked to a climax highlighting the limitations of existing AI. He called this 'present day AI cannot get out of the domain'. Domain refers here to the *training set*; any AI effort is initialized by feeding a huge collection of existing patterns to the neural network. AI really means that in no time the network finds out how the next pattern fits in.

But the trouble is that it will not do anything with a pattern that it has not seen before. This is the meaning of 'staying inside the domain'. Now it comes.

By definition a good story is a *new* pattern, it is in one or the other crucial way, different from any existing pattern. This text passage is itself an example. In all of the internet you may not find a pattern referring at the same time to AI and 'Sapiens' in the specific way that I just presented.

There is a huge training set associated with AI, and one can load up the neural net with a huge collection of original stories (e.g. using Encyclopedia Britannica as training set). But when there is original thought at work, AI is just blind because it is not part of the training set. Surely, by telling it to the computer it does become part of the training set, but will it figure it out all by itself? The computer professionals have been chasing such acts for years but up to now to no avail. It appears to be empirical fact that the present day *AI algorithms are incapable of story telling in the guise of Harari!*

The take-home message is that the capacity of our human brain to construct stories is quite mysterious. It is an old dictum by Feynman that you really understand something when you can reconstruct it by engineering it. In this regard the computer scientists hit a brick wall: for reasons that are completely in the dark, our fabulous information processing machines are inferior even to human toddlers when it comes to 'original thinking', whatever it means. Of course, humans directly understand what is meant, but it appears to be impossible to capture it 'in silicon'. Perhaps it is the case that from the very moment that mankind started its story-telling habit, we have been intrigued by the mystery called consciousness. But the complete failure by the biggest supercomputers to capture any of it, despite a concerted effort by very competent computer scientists over nearly a century leaves no room for doubt. There is something very big behind story telling and we have presently not the faintest clue what it is.

6.3 A simple question due to Plato

But we are perhaps reinventing a wheel. Becoming older, I am getting increasingly intrigued and impressed by the classic Greek philosophers—it seems that they have some deep insight in the offering, even in the present day regarding anything that really matters, roughly 2.5 millennia after they made it up. It is also the birth place of mathematics with giants like Pythagoras, Archimedes, and Euclid. Since mathematics was a fresh affair in this classic Greek culture, it took a prominent role in their philosophical arena. From a contemporary viewpoint, it appears that the philosopher Plato got it to a point with his metaphysical 'Platonic realism'. In essence, it claims that next to our daily material world there is another reality stitched together from 'Forms', perfect mathematical-like objects capturing the non-physical essences of things of which physical things are imperfect imitations. I remember well the example quoted by the professor when I took the history of philosophy course as a student: a straight line in mathematics is perfectly straight, while the straightest line in the physical universe found on a state-of-the-art chip will still jitter, be it on a sub-nanometre scale.

Anybody who has got a real grip on mathematics will agree that the more you learn, the more one gets in the grip of the seemingly endless, austere beauty of the subject. In my subjective experience that I share with many others, it is an aesthetical experience reminiscent of the way that the beauty of music can please our brains. But there is more to it, and the bulk of my present text is a case in point. This was completely anchored in the, yet again, beautiful equations forming the foundation of the grand theories of physics. But the mathematics acts as a telescope, the 'mathematical eyeglass', capable of stretching our 'intuition' into realms that are completely detached from our animal experiences. But intuition really means our story-telling capacity! I am of the strong opinion that mathematics is the asymptote, if you wish the climax of Harari's magical stories.

But there is even more. Thanks to maths, mankind invented physics and the equations of physics spawned in turn the discipline of *engineering*. All the machinery that elevated our race from hunters living in caves, to twenty-first century comfort is without exception rooted in the expanded horizons seen through the mathematical eyeglass. Engineers have to learn a lot of maths during their education—mechanical engineers employ mechanics, but also for instance elasticity and hydrodynamics, resting on quite heavy calculus. Electrical engineers have to cope with fully fledged Maxwell electrodynamics, the 'first' field theory with its vector calculus, but also with a lot of quantum mechanics and so forth. Imagine, where would your cell phone be without all the help by mathematics? In a way, the gadget is 'made from equations'.

Now it comes. It is a secret in the contemporary mathematics and theoretical physics communities. It is called 'Platonic' and it revolves around a very simple question: *is mathematics discovered or invented?* It turns out that the vast majority of pure mathematicians are of the opinion that it is *discovered*. It is a practical, daily life research strategy affair. Like in the empirical sciences, the mathematicians have the perception that they are searching around in a landscape that is just there, waiting to be explored. It reveals a stunning 'fabric' of unexpected relations that just show themselves, and it appears outrageous to blame this on humans that are inventing it. Take the Platonic solids (Fig. 6.1); their perfection is eternal, not requiring interference by human action.

But when maths is discovered it means that there is a mathematical objective reality that lives in parallel with our material world: we are forced to take Plato's parallel-universe-of-forms seriously! Given the second thoughts in the above, it seems than also reasonable to associate this in turn with the *asymptote of story telling*. Let's say that all other forms of story telling are imperfect derivatives of the almighty story-telling affair called mathematics.

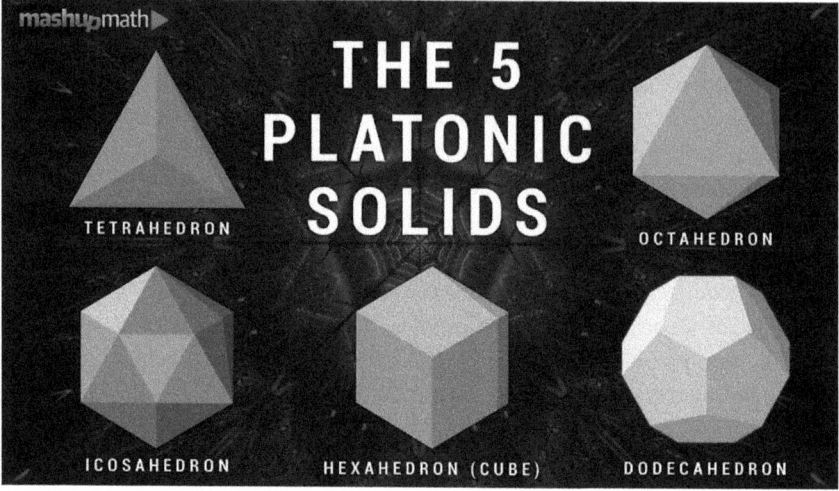

Fig. 6.1 The Platonic solids as an icon of Platonic perfection.

But this appears to mean that attempts to capture this in silicon may be futile. There is just Plato's parallel universe and 'intelligence' means as much as the capacity to view what is happening in this other world. It is as if mankind got equipped some 70,000 years ago with a 'telephone line' connecting its animal brains to this theatre of real knowledge. I find this very romantic and therefore to be considered with maximal scepticism. If true, we are a chosen people with a brain capable of sheer magic that makes us into something much more than phenomena that can be explained completely in terms of the established wisdoms of physics, chemistry, and biology.

In this regard, it seems that somehow buried in the subconscious of my professional community there is a shared sense that we may miss something essential in our understanding of specifically the workings of our brains. Among biologists one finds some quite rabid atheists arguing that it is all about molecules and so forth, while Jesus should shut up. But in a way biology is applied physics, and I have just been arguing that on this more fundamental level there may be crucial matters that are not at all captured by the fundamental equations that eventually rule also the world of molecules. The bottom line is that there appears to be quite some more room for spirituality and even religion in the physics community as compared to the adjacent empirical sciences.

I find it a bit scary. It may well be that we are still completely missing the point regarding the essence of reality. The 'Platonic view' and everything else I have been discussing in this context is no more than a metaphor. It may be very different from anything that we can presently imagine—yet another helpful metaphor refers to the Vikings, being in the early middle ages technologically the most sophisticated culture in Europe, explaining thunderstorms in terms of very muscular warriors quarrelling with each other. Think instead of thunderstorms, of 'consciousness' and instead of the warriors the Schrödinger and Einstein equations.

6.4 Quantum physics, consciousness, and story telling

Let's turn back to the main theme of this text—time, causal versus non-causal. We surely discern a hierarchy. Our time, which is the time of GR, 'supports' causality. But causality is at the least a necessary condition for *information processing*. One may take this a step further to a presently fashionable notion: 'its from (qu) bits'. This speculates that reality is really about a flow of information. It is metaphorically like the motion picture 'The Matrix' where all of reality is just a software code. This may all make sense.

However, when any of the Platonic view makes any sense at all, it seems that we have to discriminate between *different* types of information. On the one hand, the 'bits' making up the information that can be tracked by computers, the 'inside the domain' AI affair. But the Platonic view then insists that there is the 'higher quality' story-telling type of information, that is sensed by the human brain coming to a climax in the big book of mathematics.

Departing from the 'eternal tranquility' of the unitary time evolution, it is a proven fact that the wavefunction collapse is a necessary condition for a computation to happen—the readout of the quantum computer. But this information processing alludes entirely to the non story-telling way that computers compute. However, now we realize that there may be the 'higher' story-telling way of processing (creating) information. But in turn it appears that this requires the intervention of human *consciousness*. You may see it coming: isn't it the implication that somehow our consciousness has to play a role in the collapse?

I was picking on the old ideas of 'subjective' wavefunction collapse, an affair going back to the early days of quantum mechanics. Specifically, the case was made by von Neumann and Wigner that consciousness is necessary for the completion of the quantum measurement process. This idea then propagated all the way to the 1960s Californian hippies marrying it with Taoism and so forth [15].

But we now understand that insofar as the blunt pattern recognition of AI is involved, there is really no difference between computers and brains. There is just nothing special regarding the information-processing capacity of the human brain, so why invoke it when one addresses the workings of a 'standard' quantum computer that is like a normal computer with its failing story-telling capacity. But could it be that the potentiality of story-telling information processing is somehow latently present in the unitary evolution, in the same guise that computer-computing departs from the unitary sector? Metaphorically, as if we have to add special parametric time dependence to the Hamiltonian followed by a special collapse requiring a conscious observer to give rise to story-telling information processing in the macroscopic world, that now somehow includes Plato's parallel universe of forms?

Bibliography

1. See my collection of editorial *Nature* and *Science* commentaries: https://www.lorentz.leiden univ.nl/zaanen/wordpress/news-and-views/comments/
2. This paraphrases a famous political quote: https://politicaldictionary.com/words/its-the-economy-stupid/
3. e.g. S.M. Carroll, *Spacetime and geometry* (Addison Wesley, 2004).
4. A.J. Beekman *et al.*, Dual gauge theory of quantum liquid crystals in two dimensions, *Phys. Rep.* **683**, 1 (2017); https://arxiv.org/abs/1703.03157.
5. J. Zaanen, F. Balm, and A. Beekman, Crystal Gravity, *SciPost Phys.* **13**, 039 (2022); https://www.scipost.org/SciPostPhys.13.2.039/pdf
6. B. Keimer, S. A. Kivelson, S. Uchida and J. Zaanen, From quantum matter to high Tc superconductivity in copper oxides, *Nature* **518**, 179 (2015).
7. J. Zaanen, *Lectures on quantum supreme matter*, https://arxiv.org/pdf/2110.00961.pdf (2021).
8. J. Zaanen, Y.W. Sun, Y. Liu, and K. Schalm, *Holographic duality in condensed matter physics* (Cambridge University Press, 2015).
9. S. Sachdev, Quantum phase transitions (Cambridge University Press, 2011).
10. These lecture notes were instrumental in spreading this word in the late twentieth century; J.B. Kogut, An introduction to lattice gauge theory and spin systems, *Rev. Mod. Phys.* **51**, 659 (1979).
11. A.A. Abrikosov, L.P. Gorkov, and I.E. Dzyaloshinski, *Methods of quantum field theory in statistical physics* (Dover, New York, 1963).
12. D.M. Ceperley, Path integrals in the theory of condensed helium, *Rev. Mod. Phys.* **67**, 279 (1995).
13. S. Borsanyi *et al.*, Ab initio calculation of the neutron-proton mass difference, *Science* **347**, 1452 (2015). https://en.wikipedia.org/wiki/Born_rule.
14. G. Zukav, *The dancing Wu-Li masters: an overview of the new physics* (Morrow, New York, 1979); see also F. Capra, *The Tao of physics* (Shambala Pub., 1975).
15. M. Srednicki, Chaos and quantum thermalization, *Phys. Rev. E* **50**, 888 (1994); J.M.Deutsch, Quantum statistical mechanics in a closed system, *Phys. Rev. A* **43**, 2046 (1991).
16. L. D'Alessio, Y. Kafri, A. Polkovnikov, and M. Rigol, From quantum chaos and Eigenstate Thermalization to statistical mechanics and thermodynamics, *Adv. Phys.*, **65**, 239 (2016), https://arxiv.org/abs/1509.06411.
17. These ideas were thoroughly explained in the book R. Penrose, *The emperor's new mind: concerning computers, minds and the laws of physics* (Oxford Univ. Press, New York, 1989). But he also pushed here the highly controversial idea that the human brain works like a quantum computer.
18. L. Diosi, Models for universal reduction of macroscopic quantum fluctuations, *Phys. Rev. A* **40**, 1165 (1989).
19. T.H. Oosterkamp and J. Zaanen, A clock containing a massive object in a superposition: what makes Penrosian wavefunction collapse tick?, https://arxiv.org/pdf/1401.0176.pdf (2014).
20. J. van Wezel, T.H. Oosterkamp and J. Zaanen, Towards an experimental test of gravity induced quantum state reduction, *Phil. Mag.* **88**, 1005 (2008); https://arxiv.org/pdf/0706.3976.pdf.
21. W.C. Yang, C.Y. Xia, H.B. Zheng, M.Tsubota and J. Zaanen, Motion of a superfluid vortex according to holographic quantum dissipation, *Phys. Rev. B* **107**, 144511 (2023).
22. Y.N. Harari, *Sapiens: a brief history of humankind* (Harper-Collins, 2015).

Index